The contents may well be considered to be unusual, and perhaps outstanding, in that their main object is to draw the attention of those directly interested in matters such as the construction and equipment of dwellings and other occupied buildings to the importance of physiological requirements as affected by construction and engineering. Although the physiological needs of man, both whilst resting and whilst working are well established and documented, the author hopes to make it plain that much valuable information, from most reliable sources, is at present either unknown or not taken into account by those to whom it may be of considerable value.

Indications are given as to how physiological needs in any environment may be measured instrumentally. Where the need is shown to exist, indications are given as to what control of physical conditions are required, and moreover how the needed modifications may be brought about. Methods of house warming that give alternatives to the use of solid fuels are described, with their relevancy to the probable increase in the use of atomic energy.

INTERNATIONAL SERIES OF MONOGRAPHS IN
HEATING, VENTILATION AND REFRIGERATION

GENERAL EDITORS: N. S. BILLINGTON and E. OWER

VOLUME 4

THE CONTROL OF INDOOR CLIMATE

OTHER TITLES IN THE SERIES IN
HEATING, VENTILATION AND REFRIGERATION

THE CONTROL OF INDOOR CLIMATE

BY

T. C. ANGUS

D.F.C., D.Sc.(Lond.), M.I.E.E., M.I.H.V.E., C.Eng.

*Formerly Senior Lecturer, Department of Applied Physiology,
London School of Hygiene and Tropical Medicine*

WITH AN INTRODUCTION BY

G. P. CROWDEN

O.B.E., T.D., D.Sc.(Lond.), M.R.C.P., M.R.C.S.

*Emeritus Professor of Applied Physiology,
University of London, London School of Hygiene and Tropical Medicine*

1966
THE QUEEN'S AWARD
TO INDUSTRY 1966

PERGAMON PRESS

OXFORD · LONDON · EDINBURGH · NEW YORK
TORONTO · SYDNEY · PARIS · BRAUNSCHWEIG

Pergamon Press Ltd., Headington Hill Hall, Oxford
4 & 5 Fitzroy Square, London W.1

Pergamon Press (Scotland) Ltd., 2 & 3 Teviot Place, Edinburgh 1

Pergamon Press Inc., 44–01 21st Street, Long Island City, New York 11101

Pergamon of Canada Ltd., 207 Queen's Quay West, Toronto 1

Pergamon Press (Aust.) Pty. Ltd., 19a Boundary Street, Rushcutters Bay,
N.S.W. 2011, Australia

Pergamon Press S.A.R.L., 24 rue des Écoles, Paris 5^e

Vieweg & Sohn GmbH, Burgplatz 1, Braunschweig

First edition 1968

Library of Congress Catalog Card No. 68-21381

PRINTED IN GREAT BRITAIN BY A. WHEATON & CO., EXETER
08 012729 0

To

Professor "Phil" Drinker

Engineer, Physiologist and Friend

CONTENTS

been possible to find evidence of the interiors of tropical houses in which the humidity, apart from dry-bulb temperature, has been so far reduced by air conditioning as to give rise to complaints. Examples of observations in tropical lands of temperatures and humidities recorded, with the recorded subjective sensations of the occupants concerned.

EDITORS' PREFACE

MODERN industrial civilization depends for its existence on man's control of his environment. Simple comfort requires that in most parts of the world buildings must be artificially heated or cooled during some part of the year. Rising standards of living have made people intolerant of the conditions of yesteryear in factories, offices and the home, and manufacturing processes themselves are requiring ever closer control of environment. Present-day air travel would be impossible without the air conditioning of aircraft.

Heating and air conditioning, then, has an essential contribution to make to the life of everyone—in the home, at work, while travelling or during recreation. These engineering services can account for between one-tenth and one half of the total cost of a building, depending on their complexity and sophistication. They require expert design; and the number of skilled personnel is, almost everywhere, too small.

These, then, are the justifications for a series of textbooks dealing with the design of heating and air conditioning plant and equipment. The series is planned to include the following subjects:

Basic principles of heating and ventilating
Heating and cooling load calculation
Heating and hot-water supply
Ventilation and air conditioning of buildings
Industrial ventilation
Fuels and boilerhouse practice
Heat and mass transfer
Fans
Dust and air cleaning
Refrigeration technology

Each volume in the series is complete and self-contained in so far as the technical and practical engineering applications of its main theme are concerned, but for a more detailed discussion of the underlying principles of certain subsidiary subjects and for derivation of the formulae and equations quoted reference to the other volumes may be necessary. For example, heat transfer formulae must be quoted and used in more than one of the books but their derivation is given in all necessary detail in the specialist volume on heat transfer. Similarly for heating and cooling load calculations which concern not only ventilation and heating but also refrigeration. This treatment has allowed more detailed consideration of the subject than is possible in an omnibus volume of manageable size.

Another book that should be consulted when more detail is required is *The Measurement of Air Flow* by Ower and Pankhurst (Pergamon Press, 1966), which does not form part of the series because it covers a considerably wider area.

The authors have taken as their starting-point a basic training in general engineering such as may be acquired during the first years of apprenticeship. On this foundation, the specialist treatment is built and carried to a level approximating to that of a first degree. The graduate engineer or physicist who wishes to enter this field will also find the series useful, since he is introduced to new disciplines (for example, human physiology or climatology) and new applications of his fundamental knowledge, while some parts of his undergraduate course work are taken to much greater depth. Throughout the whole series, the practical applications are stressed.

The volumes do not pretend to cover the whole range of problems encountered in design, though a student who has mastered the basic principles embodied therein should be a competent engineer capable of handling a majority of the tasks he will meet. For the rest, practical experience backed by further study of more advanced texts will be essential.

N.S.B.
E.O.

AUTHOR'S PREFACE

WHEN I began to plan this book the magnitude and the extent of the material to be covered were, at first, hard to realize. One of my greatest difficulties has been to decide what to ignore whilst attempting to give some, though at best a very inadequate, treatment of much excellent material. Early it had been suggested to me that the approach should be made rather from the engineering and architectural sides than from that of the medical. No one can dispute the need for more research, and the efforts and the moneys now being devoted to research are very great indeed. But all conversant with the subject agree that, except in the fields of medicine and surgery, more urgent than the call for more research is the need for more applications of the accepted and proven findings and results of pioneers and investigators. Particularly is this true of that "poor relation" of preventive medicine—hygiene engineering.

For example: There are two simple and cheaply instrumented measurements on which are based two scales, by which the physical quantities known to control human efficiency and comfort, as affected by environment, can be evaluated. It is not too much to say that the existence of these two measurements is generally unknown today.

Of necessity my dealing with historical backgrounds is brief and incomplete, but I hope to give some indication of future trends in the management of indoor climates in many lands.

The usual attitude of the European pioneers to tropical climates, not surprisingly, was one of confrontation with a powerful and implacable enemy. We have been shown that it is possible to collaborate with Nature, and to live in amity with tropical surroundings: making careful use of local materials and vegetation, whilst taking every possible advantage of terrain.

For those living in cold climates, and in view of the limited resources of natural fuels, it may not be unreasonable to assume that the utilization of atomic power, rather than heralding the end of mankind, may well mark the opening of such new chapters in the story of civilized living as appeared with the development of the steam engine. Who in the days of the first steam-operated mine pumps in Cornwall could have foreseen express trains, ocean-going liners or the lighting of all houses by electricity? In a fairly recent lecture at the Institution of Civil Engineers H.R.H. Prince Philip, Duke of Edinburgh, maintained that power is of an importance equal to that of food. Surely there are signs that if we use our many new potentials wisely, widespread hunger and want may be "on the way out". The future should be in our hands.

This book is not intended to be a textbook purporting to instruct the designers of heating, ventilating and cooling appliances, or installations. Neither is it intended to give detailed directions in the making of heating or ventilation surveys, or in the use of the various techniques and instruments needed for such surveys.

Rather, the main objects in writing are two. Firstly, to arouse, in those directly concerned, particularly users and potential users, a much wider and greater awareness of modern designs and appliances and their value if wisely used. This is considered to be particularly important for those interested in the erection and equipment of new buildings. Secondly, for our practitioners in the arts and technicalities of heating, cooling and ventilation it is hoped to give a guide to proper use of up-to-date measurements and standards of the factors affecting human efficiency and comfort as affected by indoor environment. References are given to sources of sound relevant information, which should be helpful in the application of the principles outlined in the book. In this last connection it must be realized that the items and examples cited as current practice, especially in the realm of heating, may be anything but new. The preparation of this book has covered a considerable period of time, during which the author has become aware of changes in the approaches to, and the techniques employed for, indoor heating.

So he calls to mind one of Professor Sylvanus P. Thompson's sayings in his inaugural address to freshmen engineering students: "Your textbooks are out of date before they can be published."

THOMAS C. ANGUS

Clavering, Essex

ACKNOWLEDGEMENTS

FOR permission to make extensive quotations and to reproduce figures I am particularly indebted to Mr. Maxwell Fry and Miss Jane Drew, with B. T. Batsford Ltd., and Sam Lambert. Also to Sir Frank Markham, with the Oxford University Press, as well as to John T. Appleby, with Messrs. Alfred K. A. Knopf of New York, for a similar privilege. I also wish to thank Doctors C. S. Leithead and A. T. Lind, with Cassell Ltd., Dr. Geoffrey Taylor and Professor D. B. Bradshaw.

Included in these acknowledgements are Her Majesty's Stationery Office, the Medical Research Council, the American Society of Heating, Refrigerating and Air Conditioning Engineers, the *Practitioner*, the *Health Education Journal*, the Royal Society of Tropical Medicine, W. H. Welch, with the Institution of Gas Engineers. Also the proprietors of the *Philosophical Magazine*; the *British Journal of Industrial Medicine*, with the *British Medical Journal*. And the proprietors of *Ergonomics*. I would also thank the Gas Council, the Director of the Building Research Station, the Royal Society of Tropical Medicine and Hygiene and the Harlow Development Corporation, Essex.

I am very greatly indebted to my late friend and former colleague Professor Guy Pascoe Crowden, O.B.E., who came to a sudden and quite unexpected end of a long and distinguished career within a very few days of the completion of my writing of this book.

Not only had Professor Crowden given me a very generous Introduction but he also gave me much good advice, with valuable additions and corrections. For much somewhat similar assistance I am indebted to Mr. Ernest Ower without whose influence this book would not have been written. My son, Mr. T. H. Angus, also helped me in the checking of most of my typescript.

For much valuable and very generous assistance I would record my gratitude to Mr. M. E. Oliver, of the Institution of Heating and Ventilating Engineers and his assistants in the Library. Also to Mr. V. J. Glanville, Librarian of the London School of Hygiene and Tropical Medicine and his staff.

For information concerning much new technical development, and for illustrations of examples of modern practice in heating, ventilation and instrumentation, I have to thank Colt Ventilation and Heating Ltd., Pilkington Brothers Ltd., Hoover Ltd., C. D. Casella Ltd., Redfyre Ltd., Copperad Ltd., and Robinson and Willey Ltd.

INTRODUCTION

THIS much-needed book *The Control of Indoor Climate* deals with many problems of heating, cooling and ventilation, the practical solution of which by engineers and architects must ensure that the physical environment in dwellings and workplaces satisfies requirements for health, namely conditions which do not impose harmful stress on physiological processes which enable man to maintain fitness from day to day. I was particularly pleased to be invited to write an introduction for his book by Dr. T. C. Angus, my former colleague in research and teaching in the Department of Applied Physiology at the London School of Hygiene and Tropical Medicine. Dr. Angus is exceptionally qualified as an engineer to deal with the subject of the control of environment in the interests of man, as he has worked in close association with physiologists throughout his professional career. After starting engineering at the City and Guilds (Finsbury) College and serving in the 1914–18 War in the H.A.C. and R.A.F. he became personal research assistant to the late Sir Leonard Hill, F.R.S., at the National Institute for Medical Research. He collaborated in research on the physiology of heat, the development of the kata-thermometer for assessment of the effects of ventilation and air movement on man, and carried out field studies on ventilation in factories and ships.

In 1932 Dr. Angus was appointed lecturer in the Department of Applied Physiology at the London School of Hygiene and Tropical Medicine, but before taking up his duties he held a Rockefeller Fellowship at Harvard University School of Public Health, Boston, and worked with the American pioneers, Professors Philip Drinker, T. F. Hatch and C. P. Yaglou, with whom he studied the human physiology of ventilation, dust control, and air analysis. His appointment to the academic staff of the London School of Hygiene forged a valuable and productive link between engineering practice and the medical science of physiology.

He made use of his experience in the U.S.A. in teaching postgraduate medical students taking the D.P.H. course at the School and in laboratory research on air conditioning for thermal comfort in the tropics, as well as in dealing with many problems of hygiene engineering in factories in England. During the 1939–45 War he was attached to the Ministry of Supply as ventilation engineer in factories, and on returning to the School of Hygiene he collaborated in field and laboratory research on the measurement of domestic ventilation in post-war houses for the Ministry of Works and the D.S.I.R. In 1952 he visited various institutes in Nigeria and made a special study of thermal comfort conditions in a large teaching laboratory at University College, Ibadan, where he made use of a jet-fan he had devised for increasing air movement for the comfort of operatives in factories in England. I have recorded this summary of his work on problems involving the application of physiology with engineering practice in order to convey some idea of the background of experience behind his treatment of the subject of *The Control of Indoor Climate*.

In his preface to his book Dr. Angus rightly emphasizes the urgent need for much wider application of the proved findings of research to present-day problems of hygiene engineering. In the opening chapter he reviews the historical evidence of man's early recognition of the need to provide indoor warmth in temperate and cold climates and gives examples of methods of heating some of which hold good today.

In the second chapter he deals in particular with research in the U.K. and the U.S.A. which has provided for engineers, architects, medical officers and authorities responsible for housing not only the standards of indoor climate which satisfy human requirements but also reliable instruments and methods for checking the attainment of those standards or for detecting physical factors causing discomfort or hazards to health. In Chapter 3 Dr. Angus examines the physiology of body heat balance in relation to the thermal characteristics of the domestic environment and gives many examples of field studies and findings of research. He draws particular attention to the special needs of the elderly for indoor warmth. This is a problem of great and increasing importance at the present time owing to the large number of persons over the age of 65 in the population, the figure being over $5\frac{1}{2}$ million in 1964. The housing of old people is a problem and a challenge to heating and ventilating engineers, architects and authorities responsible for social welfare.

In the later chapters of his book Dr. Angus has done well to assemble and place on record many instances of successful practice by engineers and architects in dealing with control of indoor climate in temperate and tropical countries. In this latter connection he draws particular attention to the writings of Maxwell Fry and Jane Drew, and also to reports of research on heat stress by workers in the U.S.A. and the Middle and Far East in recent years.

In industry the increasing tempo of production, new processes and automation all involve the human factor and therefore call for the urgent application of existing knowledge in order to ensure that the environment of work meets human requirements for health, thermal comfort and efficiency.

GUY P. CROWDEN

Stanmore,
June 1966

CHAPTER 1

THE NEED FOR THE CONTROL OF INDOOR CLIMATE

In his adaptability to live his life in widely separated parts of the world, man has few rivals in the animal kingdom. With the exception of the humble sparrow, it is unlikely that any other animal or bird can be imported to places with such very different climates, there to live and multiply.

MAN'S THERMAL NEEDS

The most obvious requirement of any dwelling is to enable people to rest, eat and sleep; and even in very hot climates this not infrequently entails that the inhabitants need to be guarded from cold at night. In cold and in temperate climates some means of warming houses is essential in winter. But it is not always realized that the object of lighting a fire, or the turning on of a radiator in a cold room, is not to add heat to the bodies of the inhabitants, but to control the rate and the modes of their cooling. It is a physiological necessity that a healthy person should continually be giving out heat to his surroundings. This heat arises from the burning, or metabolism, of the food taken in, much of the resulting energy of which is converted in various muscles into the mechanical work needed to walk, work, or to move the body or objects. It is true to say that the only occasions when heat should be directly added to the body of a person are after prolonged and profound chilling, such as results from long immersion in the sea, exposure in a blizzard, or when very old persons, generally indigent and poorly nourished, are found suffering from hypothermia in unheated houses. The rewarming of such sufferers may be dangerous if not carried out under medical direction.

THE HEAT OUTPUT OF NORMAL MAN

If it is assumed that the main purpose of food is to provide energy for the performance of "work", here used in its widest sense, i.e. the

1

movements essential to life, the heat generated in the body and dissipated to the body's surroundings may be considered as a wasteful by-product. In an electrical power station of usual design only some 30 to 35% of the heat energy of the fuel burned is delivered as useful electric power, to drive trains, to light and warm our houses, and to run factories. The remaining 70 to 65% of the possible energy has to be disposed of as heat, taken from the condensers in the cooling water, passed up the chimney in hot gases, and lost to the surrounding air from hot surfaces and in internal friction. There is another analogy. In a power station there are many auxiliary machines: pumps, forced-draught fans and mechanical stokers. All these auxiliaries are essential; and all consume power. Similarly, in resting or sleeping man the heart muscle must continue to take power, the lungs must not stop, and the millions of living cells that constitute the body and its organs must be in active metabolism, or they would not continue to live. This heat loss at complete rest is termed basal metabolism, and its value is very considerably less than that produced by a man when working, walking, or even sitting up. In this connection it is interesting to note that the overall thermal efficiency of a trained athlete "full out" is about 30%—much the same as that of a good, but not the most modern, electric power station.

THE NEEDS OF CIVILIZED MAN

According to Markham (1947) it has been shown conclusively that all the ancient civilizations of which we have traces developed in those parts of the world where, on account of equable climate, and with no undue difficulty in obtaining food, the struggle for existence did not entail the use of artificial protection from excessive heat or cold provided by elaborate clothing and buildings. Markham gives maps showing the sites of these origins of early communities round the world; and they show that, almost without exception, these civilizations flourished in limited areas along the comparatively narrow line of the 70°F (21·1°C) isotherm. That is where the average air temperature throughout the year does not vary significantly from this value.

As will be seen in Fig. 1.1 this equable temperature is found in northern Egypt, whence the 70°F isotherm can be traced through Palestine, Syria and southern parts of Arabia, Sumeria, with the sites of Babylon, Nineveh and other great cities, following on into

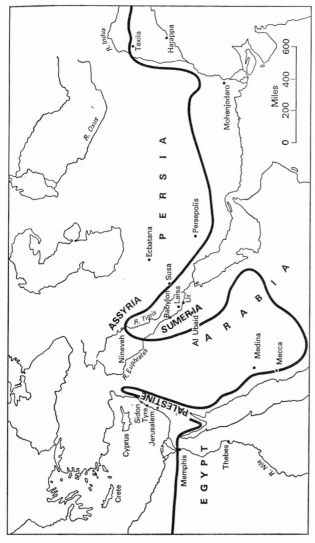

FIG. 1.1. The 70° isotherm and the sites of ancient civilization. The isotherm is shown as a line for simplicity, but it is rather a succession of areas.

northern India. In the Americas Markham shows that although the civilizations of the Aztecs, the Incas and the Mayas have left their monuments in Mexico, Yucatan and the adjoining districts, their civilizations started in Central American highlands, almost along the line of the 70°F isotherm.

There was a widely held theory, now somewhat discredited, that during historic times considerable changes of climate had taken place in Mediterranean regions. It is now believed that the only considerable change in recorded times has been that Babylonia, Persia, Egypt, and even India, at the height of their periods of glory, were moister than they are today. Markham writes: "This work seeks to determine whether or not climate and climate control influence civilization. . . . After full consideration, I am convinced that one of the basic reasons for the rise of a nation in modern times is its control over climatic conditions: that nation which has led the world, leads the world, and will lead the world, is the nation which lives in a climate, indoor and outdoor, nearest to the ideal, provided that its numbers are large enough to resist invasion by its rivals. Civilization to a great degree depends upon climate control in a good natural climate."

Progress in the provision of appliances for the warming of the dwellings of the inhabitants of lands with cold climates was probably slow. But there is ample evidence that with the expansion of the Roman Empire the control of cold and damp was practised in lands far from Italy, as witnessed by the many foundations of Roman buildings containing hypocausts that have been found in England. Markham quotes Dr. Stanley Casson to the effect that the Ionians may have been the originators of this form of central heating, and that at Ephesus the famous temple was centrally heated by means of lignite. In this connection it is interesting to note that another temple—the as-yet unfinished cathedral in Liverpool—employs a hypocaust method of heating. Here copious volumes of heated air are impelled through large air ducts, the tops of which form the floor of the cathedral.

Another and newer form of floor heating is now in course of installation in our 700 years old Salisbury Cathedral. Here the Choir and the Lady Chapel have been equipped with off-peak electric floor heating; and it is intended that the whole of the building shall be similarly treated. We are informed (Electric Floor Warming, 1966) that the reason for this innovation was for the preservation of the high parts of the cathedral structure which were suffering from the effects of frost and damp. Other great improvements that resulted were entirely fortuitous. During the exceptionally severe winter of 1962–3 both superior comfort and greater economy in operation were afforded in the Choir and the Lady Chapel of the cathedral, where the floor heating had been installed. A graph of temperature measurements made at 7 a.m. during January 1964 shows that floor surface temperatures ranged between 68°F (20·0°C) and 80°F (26·7°C); air temperatures at 8 feet above the floor were nearly constant at 53°F (11·7°C), with outside temperatures ranging between about 26° and 30°F (−3·3 and −1·1°C); and at the same time the vertical temperature gradient over the floor panels was small, the air temperature near to the inside of the roof and of the roof itself varying by only about 1°F (0·6°C) from the air temperature at head level.

Details are given in the 1966 reference of the laying of the heating elements in pre-cast tubes in the upper concrete screed, the heat insulation below these, and the permanent flooring of honey-coloured Purbeck stone. The heating is controlled by a floor thermostat and a wall-mounted air thermostat, both in series with the main time-switch, and intended normally to limit the floor surface temperature to 75°F (23·9°C) and the air temperature to 55°F (12·8°C).

Markham (1947) writes: "Apparently the Lacedaemonians were the first to discover that by passing hot air through the floors, and later through the walls of buildings, a method of warming buildings superior to the use of open fires or brazier was obtained. . . . Let us consider what this meant to the average inhabitant of an Italian city. The wealthy, of course, had their own heating arrangements, but the mass of the people could go to the Baths in the middle of the Italian winter (which at Rome meant an average January temperature of 44°F (6·7°C), assuming no great climatic change, and at Milan an average January temperature of about 34°F (1·1°C), and find there for a trifling sum, warmth and entertainment for several hours."

He concludes that it may not be beyond the bounds of possibility that, in the case of the Romans, and, later, of the Americans, such relief from the stresses of winter may have helped to produce a more energetic race.

HOUSE HEATING IN EARLY TIMES

It is probable that, apart from the elaborate buildings of the Greeks and the Romans, the warming of all buildings, both great and small, depended on open fires and braziers. Appleby (1947) describes the warming of English dwellings in the twelfth century. "Keeping warm in winter was always a problem. The houses, even the greatest, consisted only of a large, high-roofed hall, rather like a barn or a small church, in which the life of the household was carried out. Only recently had a separate bedroom for master and mistress been introduced, and the rest of the household slept on rushes on the floor. In the larger establishments food was usually prepared in a separate building. When King John had his houses at Marlborough and Ludgershall repaired in 1204, he ordered that a new kitchen be built at each house for preparing his dinner, with a "Furnace" in each one large enough to cook two or three oxen in. In the smaller houses cooking was done over the central fire in the hall. Meals were eaten off trestle tables set up for each meal. Stools and chests, with perhaps chairs of state for the lord and lady, completed the scanty furniture. The draughty halls were heated by log fires in the centre of the stone floor, the smoke was left to find its way out of an opening in the roof as best it could. Men tried to keep warm by wearing more and heavier fur-lined cloaks. People lived so much out of doors that they were hardened by exposure, and a man like King Henry I, with his rough red hands and weather-beaten face, would probably not have looked for much more comfort indoors than could be found under the shelter of a tree in the forest."

In the majority of European dwellings we may assume that in cold weather the occupants sat, or lay, around the central wood fire at distances dependent upon their relative importance. In even earlier times similar customs may have prevailed around the camp fires of primitive tribes in many lands; the camp fire giving not only warmth, but protection against wild animals and "devils". It is not unreasonable to suppose that the very real fear of the dark from which our own children so often suffer is inherited or in some way ingrained from the less secure times of primitive ancestors. There is still among us a real satisfaction in the wasteful and inefficient open fire which has been extolled by G. K. Chesterton and Mark Twain. The latter must have experienced to the full the rigours of North American winters in the course of a long and adventurous life, during which he had participated in the early gold seeking, and had piloted steamboats on the Mississippi. When success came to him, and he built his house in Hartford, Connecticut, he wished it to embody his ideal of complete bliss and comfort after hardship. This was: to be able to sit in front of a roaring fire, his feet on the mantelpiece, his famous corncob pipe alight, and, at the same time, to be able to gaze out upon the full fury of a winter storm. After some little architectural difficulty this haven after storm was achieved. So now the pilgrim to Hartford is able to sit in the chair, put his feet on the mantelpiece, and look out of the large window, the bottom of which almost rests upon the mantelpiece.

REFERENCES

APPLEBY, J. T. (1947) *John King of England.* Alfred A. Knopf, New York.
Electric Floor Warming (1966) *Heating and Ventilating Equipment News,* **9,** No. 1, 4.
MARKHAM, S. F. (1947) *Climate and the Energy of Nations.* Oxford University Press.

CHAPTER 2

THERMAL COMFORT: INSTRUMENTATION AND STANDARDS

FOR long it has been the object of research workers to arrive at some instrumental means by which to evaluate thermal sensation and its significance in respect of human comfort and efficiency. The four important physical factors concerned are: air temperature, humidity, radiated heat, and air movement. Of these it may be accepted that for cold and cool climates air temperature is the most important, and that under very hot conditions humidity may be of even greater importance. One of the earliest investigators of these problems was Haldane (1905) who showed that in hot mines the reading of the wet-bulb thermometer gave a better indication of the heat stresses of the miners than the dry-bulb temperature of the air. Leonard Hill's kata-thermometer indicated the cooling effects of ambient surroundings at temperatures below that of the exposed human skin (1919, 1920), and showed, for the first time, the very great significance of the rate of air movement over the person. Hill's wet kata-thermometer also indicated the influence of the humidity of the ambient air on the cooling of a moist, warm body, with or without air movement. Experience showed, however, that as a physical instrument the wet kata-thermometer leaves much to be desired.

Today the kata-thermometer is used mainly as an anemometer; for this purpose it is particularly useful because it will measure very low air velocities and also summate the cooling effects of eddies and irregular changes of velocity. These properties are of importance when evaluating the effects of air movements on man, particularly when indoors.† The silvered kata-thermometer, illustrated in Fig. 2.1, has the advantage that its readings are independent of radiation (see p. 24).

In an attempt to measure simultaneously the resultant physical effect of the four factors Missenard (1933) introduced his resultant thermometer. This was a small globe of thin sheet metal with a

† See also Chapter 7, pp. 78-80.

8

blackened surface into which was inserted the bulb of a glass thermo-meter. Upon the outer surface of the globe were fixed a few strips of muslin, the ends of which dipped into water. In this manner the additional cooling of the globe due to increased evaporation by reason of moving and relatively dry air was subject to measurement. The readings given by this instrument were termed resultant tempera-tures. Little use seems to have been made of this instrument, and few results of field applications appear to be available.

THE MEASUREMENT OF DRY-BULB TEMPERATURE

The dry-bulb air temperature is a factor that must be taken into account in all studies and assessments of the thermal sensations of human subjects. The temperature of a room is generally measured by means of a mercury-in-glass thermometer, usually held in a frame attached to a wall. Such an instrument, mounted at a height of about 5 or 6 feet above the floor, gives a reading of room tempera-ture considered to be sufficiently accurate for all needs, including the requirements of official standards for health and comfort. But for greater precision certain limitations of this instrument should not be overlooked. To indicate true air temperature, the thermometer bulb should be well clear of the containing frame and not too close to the wall; nor should the wall transmit conducted heat, as from a sun-warmed exterior or a contained chimney flue. Polished glass, unlike polished metal, has a highly absorptive surface for incident radiant heat. A wall-mounted thermometer directly exposed to a radiant fire or to a dull radiator of considerable size is subject to a radiation error which may be prevented in two ways: by silvering the exterior of the glass bulb or enclosing it in a layer of bright aluminium foil; or by screening it and at the same time creating a relative movement at considerable velocity between bulb and ambient air, as in a whirling hygrometer (see below).

EQUIVALENT AND EFFECTIVE TEMPERATURES

The Equivalent Temperature Scale and the Effective Temperature Scale, both of which were developed independently about 40 years ago, the former in England and the latter in the U.S.A., have the same purpose, namely to provide an easily understood measurement of physical conditions which will define an average person's thermal sensation. As already noted there were early attempts to make an instrument which would give an immediate single reading of ambient physical conditions corresponding to thermal sensation. Such an instrument might be of limited value. But today both equivalent temperature and effective temperature—the near-identity of their names is unfortunate and liable to cause confusion—are determined from the instrumental measurement of more than one physical quantity, with subsequent evaluation of the resultant by means of appropriate scales. A valuable result is that any of the factors that may be causing unsatisfactory conditions can be detected and appropriately corrected.

The Equivalent Temperature Scale

This is especially suited to the evaluation of the effects of heating appliances in any climate where such appliances are needed. It takes account of dry-bulb temperature, rate of air movement, and radiation from or to surroundings. The last of these can be either positive, as from exposure to a bright fire or radiator, or negative, as from radiation from warm exposed skin or clothing to a large cold surface.

The Effective Temperature Scale

This is especially valuable where high temperatures prevail and the evaporative cooling of the body assumes great importance. Its essential difference from the Equivalent Scale is that it takes account of humidity, i.e. it requires measurements of wet-bulb temperature as well as of dry-bulb. The effective temperature is widely used in America for the evaluation of thermal sensations.

EQUIVALENT TEMPERATURE

The equivalent temperature of a room, or of a position therein, was defined by Dufton (1929). It is that temperature of a uniform enclosure in which, in still air, the clothed human body would lose heat at the same rate as in the room or position under observation. Dufton described experiments which led to the development of the eupatheoscope,† illustrated in Fig. 2.2, an instrument giving immediate readings of equivalent temperature. It consists of a closed cylinder of thin copper sheet, 22 inches high and $7\frac{1}{2}$ inches in diameter,

---

† The original instrument was termed the eupatheostat; the eupatheoscope illustrated in Fig. 2 was a later development.

to which a specially designed thermometer is fixed. This cylinder, which Dufton described as a "sizable black body", is of dimensions more in keeping with those of a man's body than is a thermometer bulb. The blackened exterior surface of the eupatheoscope, which needs an electric power connection, is kept at a temperature of 75°F (21·1°C) thermostatically, so that the instrument's surface temperature is much that of a clothed man. This means that the eupatheoscope cannot be used where the ambient temperature exceeds this value.

It is stated that now neither Missenard's resultant thermometer nor the eupatheoscope are being manufactured.

Equivalent Temperature: Field Measurements

Bedford (1964) describes the implications of equivalent temperature and the development of its techniques.

Following and enlarging on the leads given by earlier pioneers, Bedford and colleagues (1936) conducted considerable research, making measurements of thermal conditions in factories, with simultaneous careful questioning of the operatives as to their subjective impressions of thermal sensation. Instead of employing the eupatheoscope for the measurement of equivalent temperature these investigators measured dry-bulb temperatures by means of whirling hygrometers, air velocities by means of kata-thermometers, and radiant effects by means of globe thermometers (see p. 12). From these observations the equivalent temperature may be calculated by the following formula:

$$t_{eq} = 0 \cdot 522 \, t_a + 0 \cdot 478 \, t_w - 0 \cdot 01474 \, V^{\frac{1}{2}} \, (100 - t_a)$$

where t_{eq} is the equivalent temperature in °F,

t_a is the air temperature (dry-bulb) in °F,

t_w is the mean radiant temperature in °F,

and V is the air speed in ft/min.

Figure 2.3 is an alignment chart from which equivalent temperature may be determined from the values of dry-bulb temperature, globe thermometer temperature and air velocity by joining the observed globe and dry-bulb temperatures on their respective scales by a straight line. The equivalent temperature is given by the intersection of this line with the observed air-velocity line. Figure 2.4 enables the mean radiant or black-body temperature to be determined, which is a useful factor in the consideration of the effects of radiant warming or cooling. Both of these alignment charts are due to Bedford (1936).

A globe thermometer, Fig. 2.5, is constructed by inserting an ordinary glass thermometer, with its bulb at the centre, into a copper sphere of about 6 inches diameter; the ball of a domestic water-cistern valve is suitable. Such an instrument freely suspended will take up the dry-bulb air temperature unless influenced by radiation.

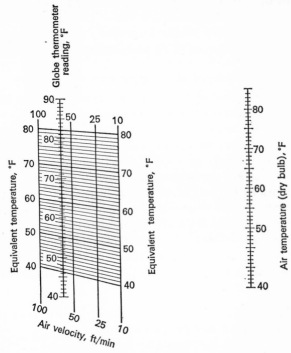

FIG. 2.3. Chart for estimation of equivalent temperature from globe thermometer readings.

The surface of the bulb is blackened; therefore if it is exposed to any source of radiation its temperature is raised, and if, at the same time, it is subjected to movement of cool air this will tend to lower the globe-thermometer reading. Similarly should the instrument be exposed to a solid surface at temperature lower than that of the ambient air the globe will lose heat by long-wave radiation to this cold surround. In this way the undesirable effects of an uninsulated thin roof can be demonstrated, in both summer and winter conditions.

Fig. 2.1. Kata-thermometer.

Fig. 2.2. Eupatheoscope.

FIG. 2.5. Globe thermometer.

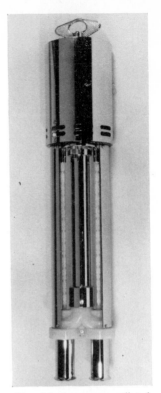

FIG. 2.8. Assmann ventilated psychrometer or hygrometer.

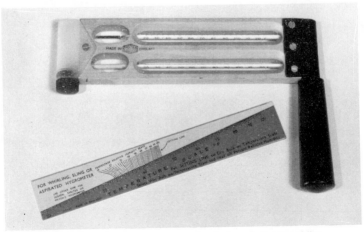

FIG. 2.9. Whirling hygrometer and rule for relative humidity.

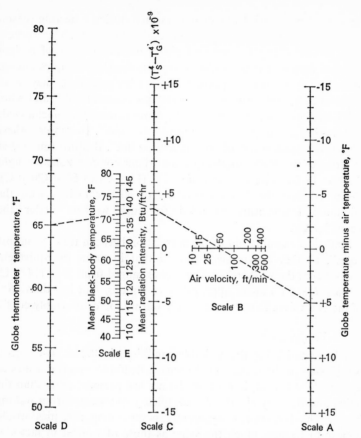

Fig. 2.4. Chart for estimation of radiant heat from globe thermometer readings. (Through the appropriate points on scales A and B a line is struck to cut scale C. From this intersection another line is drawn through the appropriate point on scale D. The intersection on scale E gives the measure of radiation.)

COMFORT ZONES

There is no "ideal temperature" for any group of people. Individuals vary greatly in physical characteristics, temperaments and adaptability. What can be done, and what has been done by a careful questioning of persons exposed to different conditions, is to establish "comfort zones" or ranges of temperature within which a large proportion of the persons concerned say that they are comfortable.

Bedford (1936) carried out such a research during a heating season in a number of factories situated in southern England. The workers were mostly young women and girls performing work of a light and fine nature, mostly whilst sitting. Bedford, in his questioning, found three groups of people: those who said that they were "comfortably cool"; those who were "quite comfortable" and who, had they the choice, would have the room temperature unchanged; and another group who were "comfortably warm". There were others who were "much too cool" or "much too hot". Bedford found that in terms of dry-bulb temperature the winter comfort zone for light workers in factories ranged from 60°F (15·6°C) to 68°F (20·0°C). He noted that the low temperature of 60°F coincides with the minimum temperature provided for such workers under the Factories Act.

Statistical examination of the correlation between these "comfort votes", as they are termed, showed that in terms of equivalent temperature the workers' comfort zones ranged from 58°F (14·4°C) to 66°F (18·9°C) under appropriate conditions. Not less than 86% of the votes were from "comfortably warm" to "comfortably cool".

Home Temperatures

When considering the desirable conditions in British dwellings it would seem to be reasonable to regard Bedford's comfort zones as applicable. After all, in ordinary homes few persons, other than the ailing and the very elderly, are completely quiescent during waking hours. The housewife's apparently never-ending activities would appear to be very much the same as those of women in factories where light work is performed; so except for children in good health, for whom too much warmth is not desirable, Bedford's comfort-zone temperature ranges should apply well in the home.

Other Criteria

Early in the war of 1939–1945 the British Government realized that a great deal of new building would be called for, and that the need for new and renovated residences would be by no means the least important. The Government accordingly set up a number of committees, which offered specialized advice on the requirements of the many various types of domestic and industrial buildings that would be renovated or newly built after the war. The Heating and Ventilating (Reconstruction) Committee of the Building Research Board of the Department of Scientific and Industrial Research (1946) published a long report from which Table 2.1 is extracted:

TABLE 2.1. Standards of Warmth in Houses
(from British Standards Code of Practice)

Room	Air temperature at 5-foot level	Relative humidity %	Equivalent temperature
Living rooms	60° to 68°F (15·5° to 20°C)	30 to 65	62° to 66°F (16·6° to 18·9°C)
Bedrooms	55° to 57°F (12·8° to 13·9°C)	30 to 65	50° to 55°F (10° to 12·8°C)
Kitchen	60°F (15·5°C)	less than 70	60°F (15·5°C)
Halls and passages	45° to 50°F (7·2° to 10°C)	30 to 65	50° to 55°F (10° to 12·8°C)
Bathrooms	55° to 57°F (12·8° to 13·9°C)	—	—
Lavatories	55° to 57°F (12·8° to 13·9°C)	—	—

It is explained in the report that the standards for dwellings recommended are based on a consensus of sound opinions and not on a widespread survey.

Room Temperatures chosen by Tenants in New Houses

Welch (1960) has given interesting particulars of the manner in which tenants living in newly built dwellings in Harlow New Town, Essex, have regulated the temperatures of their rooms by thermostatic control of the gas-fired space-heating installations. One of the aims in the planning of this new town was to avoid dust and smoke, the inhabitants being mostly young people with children. Houses built for letting varied from small two-bedroom houses of the terrace type to four-bedroomed houses with garages. A pilot installation of gas space heating and hot-water supply was installed and observations were made for six or eight months after August 1959.

Table 2.2 shows some of the findings. Out of 13 families only one said that they missed an open fire, but 12 used radiant electric fires in addition to the space heating. All but one were satisfied with the running cost. Most of the occupants opted for room temperatures of 68°F (20°C) or 70°F (21·1°C).

Table 2.2. Customers' Observations on Complete Warm Air and Hot-water Service

Customer no.	No. in family Child	No. in family Adults	Whether home all day	Age assessment of customer	Roomstat setting °F	Do you miss open fire?	Are you satisfied with running cost?	Are you satisfied with hot water and heating service?	Other forms of supplementary heating used. N.B.—All electric fires are radiant unless otherwise stated	Customer's remarks
111	1	2	Yes	Middle Aged	70	No	Yes	Yes	Electric fire 1 kW	Electric fire used occasionally in bedrooms.
112	2	2	Yes	Young	70	No	Yes	Yes	—	Satisfied with both hot water and space heating.
113	2	2	Yes	Middle Aged	70	Yes	Yes	Yes	Electric fire 2 kW Oil heater	House keeps cleaner without open fire.
115	1	2	Yes	Young	65	No	No	Yes	Electric fire 2 kW	Uses electric fire frequently and does not think it expensive. Agrees heating efficient.
116	1	2	Yes	Young	68	No	Yes	Yes	Electric fire 2 kW	Electric fire used in conjunction, mainly in evenings.
117	–	3	Yes	Middle Aged	78	No	Yes	Yes	Electric fire 2 kW	House keeps cleaner without open fire.
118	–	2	Yes	Young	70	No	Yes	Yes	Electric fire 1 kW	Electric fire used only in bedroom. Thinks heating excellent.
133	–	2	Yes	Young	70	No	No	Yes	Electric fire 2 kW	Thinks running costs high but agrees value for money.
134	–	2	Yes	Young	68	No	Yes	Yes	Electric fire 2 kW	Electric fire used to warm bedrooms.
135	1	2	Yes	Young	68	No	Yes	Yes	Electric fire 1 kW	Electric fire used to warm bedrooms.
136	–	4	Yes	Middle Aged	70	No	Yes	Yes	Electric fire 2 kW Oil radiant heater	House is much cleaner. Wonderful hot water.
137	3	2	Yes	Middle Aged	64	No	Yes	Yes	Electric fire 2 kW	Electric fire used only when sitting in one place for long period, e.g., watching TV. Note: Low temperature setting.
138	1	2	Yes	Young	70	No	Yes	Yes	2 electric fires 1 kW	Very satisfied with hot water service.

Increasing Indoor Temperatures

According to Markham (1947): "Fifty years ago, 60°F (15·5°C) was held to be the ideal winter temperature. By 1920 the British ideal was about 64°F (17·8°C) this last being more generally accepted for government buildings, offices and cinemas." One cinema administrator, a few years ago, stated that townsmen preferred a slightly higher temperature than country folk. Possibly these changes are coming about on account of a growing habit amongst the men of this country to wear thinner, and less, underclothing in winter than formerly. Markham continues: "In the United States the old ideal of 70°F (21·1°C) is being replaced by one of between 72°F (22·2°C) and 75°F (23·9°C). There is obviously a point in this where the ideal becomes a 'comfort' ideal instead of an 'energy' ideal, and it may be that the best temperatures and humidities are not those in which one feels perfectly comfortable but those in which one feels slightly cool, for coolness is a prime essential to the physical work (including typewriting), without which all mental effort becomes, as the Arab's, mere conversational speculation, barren in result."

The Overriding Importance of House Warming (*Markham, 1947*)

"No form of climate control, whether based on coal, gas, electricity, wood, oil or any other material, can be considered effective from the point of view of national energy unless it is cheap enough to be within the reach of the poorest. A given country may have an ideal climate for ten months of the year; if its means of controlling two months of bad climate are restricted to the wealthy owing to the high local cost of fuel, or even of clothing, it will not be so well off as a country with a worse climate and cheaper supplies. Central Chile has one of the best climates in the world, but absence of fuels and of cheap textiles leaves it worse off than, say, Scotland, which has a far inferior climate. . . . Of the warming and drying methods, the coal fire, with its supplement of gas and central heating, is still supreme,† while the use of electrical and oil heating devices is gradually extending. Therefore any country which has vast supplies of coal, oil, or can easily produce electricity, has means of providing its inhabitants with cheap methods of warming and drying their houses."

† This was written in 1947, but it is still true that the coal fuel is extensively used, particularly in the U.K.

EFFECTIVE TEMPERATURE

The Effective Temperature Scale was developed in the United States of America (Houghton and Yaglou, 1923; Yaglou, 1927). This scale stresses the importance of humidity in influencing human thermal sensation; this is particularly true at the higher temperatures, such as are found in certain industries, and in the tropics, where the physiological cooling of the body is so largely dependent upon sweating.

Original Researches

In these researches a number of "judges", trained observers, had to pass to and fro between two carefully conditioned rooms. Two scales of effective temperature were constructed: the "basic", for men stripped to the waist, and the "normal" for men wearing the usual light clothing customary in the United States. The originators of these scales were careful to point out that they should not be considered to be universally applicable because they were constructed from American nationals, wholly or partly clothed.

In one of the air-conditioned test rooms the air was kept as still as possible and fully saturated. In the other room the temperature and the humidity could be varied at will as the experiments proceeded. A large range of temperatures and humidities was explored, the method being to vary the dry-bulb temperature, the humidity and the air velocity in the variable room until the judges in moving from this room to the static room could declare that their sensations of heat or cold were identical in both rooms. With practice it was found that agreement among a number of subjects could be obtained to within one degree Fahrenheit. The judges were instructed to ignore, as far as possible every sensation that was not purely thermal.

To determine the normal effective temperature from the chart shown in Fig. 2.6 a line is drawn to connect the dry-bulb (d.b.) temperature with the wet-bulb (w.b.) temperature, on their appropriate scales.† Such a line cuts through a family of curves representing the recorded air velocities from 20 ‡ to 1500 ft/min. The point at which the d.b.–w.b. line cuts the air velocity curve representing the recorded velocity indicates the normal effective temperature. Inspection of Fig. 2.6 shows that, since the air velocity lines incline towards

---

† For use of globe-thermometer reading, see page 21.
‡ A speed of 20 ft/min can be taken as representing "still" air; it is difficult to avoid movement of this order even in a closed room.

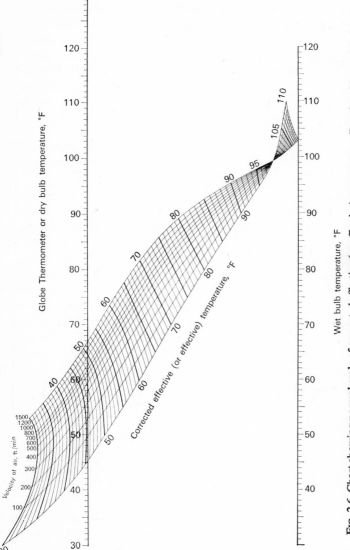

FIG. 2.6. Chart showing normal scale of corrected effective (or effective) temperature. Reprinted by permission from ASHRAE *Guide and Data Book*, 1965.

the wet bulb as temperatures rise, humidity exercises a greater influence on thermal sensation at high temperatures than at low: at low temperature, when the skin is usually dry, dry-bulb temperature is of major importance. More recent researches have resulted

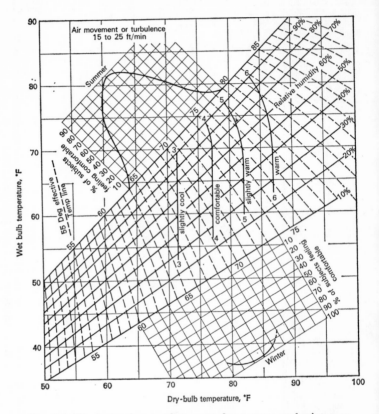

FIG. 2.7. Chart showing comfort zones for summer and winter, as experienced in North America in terms of effective temperature. Reprinted by permission from ASHRAE *Guide and Data Book*, 1965.

(Fig. 2.7) in which lines of equal effective temperature are plotted on a hygrometric chart against dry-bulb and wet-bulb temperature with lines of relative humidity. In addition there have been added comfort zones for both summer and winter, in terms of effective temperature; these have been derived from the proportions of votes given by numbers of subjects when exposed to different thermal environments. In summer the highest proportion of comfort votes was found to be at an effective temperature of 71°F (21·7°C). The corresponding most popular indoor temperature for heated rooms in winter was at the effective temperature of 67·5°F (19·7°C).

Example. Consider a reasonably hot summer day in an American house. The dry-bulb temperature is 80°F (26·7°C) with relative humidity 44%, the wet-bulb then reads 65°F (18·3°C). Referring to Fig. 2.6 it will be seen that with an air velocity of 20 ft/min a random velocity that commonly occurs in closed rooms, the effective temperature will be 73°F (22·8°C). Should this be found to be unduly warm some physiological cooling may be induced by starting mechanical fanning. If this is sufficient to produce an air velocity or turbulence of 200 ft/min where persons are situated, Fig. 2.6 shows that the effective temperature will be lowered 2°F to 71°F (21·7°C): which has been shown to be that considered to be the most suitable by the votes of the experimental subjects (see Fig. 2.7).

Effective Temperature: Modifications

After much experience subsequent to the original researches, it was realized that completely still and saturated air is a condition too artificial on which to base any valid criterion of sensations: these can only be most unpleasant, irrespective of heat or cold. The revised effective temperature as adopted by the American Society of Heating, Refrigerating and Air Conditioning Engineers (whose *Guide and Data Book* should be consulted) is now based on comparisons made by the judges with thermal sensations in a static room with completely saturated air which is moved at a constant low air velocity instead of being nominally quite motionless. It is between such a "static" room and another room, in which the three physical factors are varied at will, that the judges pass until thermal equality is agreed.

Radiation Effects. Originally, the air and the walls of the test rooms were at the same temperatures, so that the Effective Temperature Scales made no allowance for radiation effects. Bedford (1936) showed that a simple and practical addition can be made to the Effective Temperature Scales by substituting (see Fig. 2.6) the temperature as shown by a globe thermometer for the dry-bulb air temperature. The effective temperature thus determined is called the corrected effective temperature (C.E.T.).

Effective Temperature—Measurements

(*a*) *Temperatures*. The two most important quantities needed for the determination of effective temperature are the dry-bulb and the wet-bulb temperatures, and the most suitable instrument which gives their simultaneous values is a "ventilated hygrometer". From its reading the relative humidity, the vapour pressure, and other hygrometric properties of the air are easily obtained. When thermometers, one with a dry bulb and the other with a wet bulb, are exposed to air, unless the air is completely saturated with water vapour (a condition rarely found in practice) the wet-bulb instrument will give a lower reading than the dry-bulb. This "wet-bulb depression", as it is termed, by application of well-known tables or charts, enables the relative humidity of the air and other hygrometric properties to be determined. Such data are by no means absolute measurements of the quantity of vapour present, but have been determined by years of careful experiment. In connection with the use of wet-bulb temperature rather than relative humidity for the determination of effective temperature it may be noted that the total heat of the air— made up of its dry heat content and the latent heat of its water vapour—is directly related to the wet-bulb reading.

For a great many years air humidity has been measured by Mason's hygrometer, an instrument containing two similar thermometers; the bulb of one is dry and that of the other is kept wetted by distilled water from a covering of thin muslin surrounding it, the extremity of this cover dipping into a container. Under the Factories Act the indications of Mason's hygrometer is still valid as an index of allowable humidity in factories where processes demand abnormally humid air. For much meteorological work such thermometers are still used, but they are then placed in Stephenson screens, which, although protecting the instruments from sun and rain, allow free movement due to natural causes over the thermometers. A disadvantage of this type of instrument is that unless relative motion is ensured a stationary wet-bulb is surrounded by a thin layer of partly saturated air. The lowest or true wet-bulb temperature is determined when the air is passed over the wet-bulb at a velocity of about 900 ft/min. In view of the importance of obtaining a true wet-bulb temperature, the use of ventilated hygrometers is advised for determination of effective temperature.

In the Assmann hygrometer, shown in Fig. 2.8, the wet- and dry-bulb thermometers are held side by side in separate tubes. At the top of the instrument a small centrifugal fan, operated by an electric motor, draws air at sufficiently high velocity to ensure a correct wet-bulb reading. Both the bulbs are protected from the effects of radiation by being enclosed in concentric tubes of polished metal. In another form of this instrument the fan is driven by a clockwork mechanism, so that an electrical supply is unnecessary.

A less expensive form of ventilated hygrometer is the whirling, or sling hygrometer (Fig. 2.9). In this instrument the two thermometers have their bulbs exposed at the end on a small hinged frame which may be whirled by hand on a spindle in the handle shown at the top of the photograph. Here the wet bulb is kept moist by distilled water supplied by a wick that enters a small container. Readings of wet- and dry-bulb temperatures are taken after the instrument has been whirled rapidly for a few seconds. This operation should be repeated at least once, to be sure that the lowest wet-bulb temperature is obtained; particularly if the instrument has been brought in from warmer surroundings. An objection to the use of this instrument is that the whirling operation, with the quick stopping and reading of the wet-bulb, can be tiring on an extensive temperature survey.

In both of these ventilated hygrometers the dry-bulb temperatures indicated will be true: in the Assmann this is secured by the screens, and in the whirling instrument by the forced convection of the moving air which should be sufficient to overcome any ordinary radiation effect.

(b) *Radiation Effects.* The globe thermometer has been described on page 12. Should this be employed in the determination of either corrected effective temperature or equivalent temperature, the procedure is simply to leave it for 15 or 20 minutes before the reading is taken.

(c) *Air Velocity.* For comfort or physiological determinations, an instrument that sums up the effects of eddies and air turbulence is likely to be of more value than the more usual form of anemometer which measures the speeds of unidirectional air currents. The originators of effective temperature recommended the kata-thermometer whose readings depend on the convective cooling, and evaporation is affected by the total air movement including random currents in various directions. Varieties of kata-thermometers can be obtained.

Figure 2.1 shows a silvered kata; the silvered bulb renders the instrument immune to radiation effects, and so gives accurate velocity measurements unaffected by hot or cold solid surrounds.

REFERENCES

BEDFORD, T. (1964) *Basic Principles of Ventilation and Heating.* H. K. Lewis, Ltd., London.

BEDFORD, T. (1936) The Warmth Factor in Comfort at Work. Industrial Health Res. Board, Rep. No. 76. H.M.S.O., London.

DUFTON, A. F. (1929) The Eupatheostat. *J. Sci. Instr.* **6**, 249.

Guide and Data Book (1965–6) American Society of Heating, Refrigerating and Air Conditioning Engineers, New York.

HALDANE, J. S. (1905) The Influence of High Air Temperatures. *J. Hyg. (Camb.)* **5**, 495.

HILL, L. (1919, 1920) *The Science of Ventilation and Open Air Treatment,* Parts 1 and 2. M.R.C. Special Reports, No. 32, 1919, and No. 52, 1920. H.M.S.O., London.

HOUGHTEN, F. C. and YAGLOGLU, C. P.† (1923) Determining Lines of Equal Comfort. *Trans. Amer. Soc. Heat Vent. Engrs.* **29**, 163.

MARKHAM, S. F. (1947) *Climate and the Energy of Nations.* Oxford University Press.

MISSENARD, F.-A. (1933) *Etude Physiologique et Technique de la Ventilation.* Librairie de l'Enseignement Technique, Paris.

Post-War Building Studies, Heating and Ventilation of Buildings (1945), No. 19. Ministry of Works, H.M.S.O., London.

WELCH, W. H. (1960) Gas and Coke Space and Water Heating Services in Houses and Flats, Harlow New Town. Public Health Works and Municipal Services Congress, Session 1960, Nov. 14th.

YAGLOU, C. P. (1927) Temperature Humidity and Air Movement in Industries: the Effective Temperature Index. *J. Indust. Hyg.* **9**, 297.

† Yagloglu later changed his name to Yaglou.

CHAPTER 3

PERSONAL WARMING; COOLING, VENTILATION; PHYSIOLOGICAL IMPLICATIONS

It is not possible in practice to separate the considerations of the heat exchanges between men and their surroundings from those of ventilation. Crowden and Angus (1947) summarized some of the essentials of thermal exchange and ventilation for men at rest in the following passage:

"An adult sitting resting, for instance, breathes 15 to 20 cubic feet of air per hour. Of far greater significance in determining ventilation requirements, however, is the fact that the metabolic processes which respiration renders possible lead to the production of some 100 kilocalories or 400 British Thermal Units of heat per hour. The maintenance of body temperature within the narrow range of normality around 98·4°F (37°C) demands the continuous loss of this quantity of heat to the environment or surroundings. It has been shown that a thermally comfortable adult wearing ordinary clothes, sitting in a room in which the air is practically still, and at a temperature of 60°F (15·6°C) dry bulb, and 50% relative humidity, loses these 400 Btu substantially as follows:

46% or 184 Btu by radiation to the solid surroundings per hour.

30% or 120 Btu by convection to the air.

24% or 96 Btu by evaporation of moisture from lungs and skin.

The loss of 120 Btu by convection is of interest in connection with the study of ventilation requirements, for it can be readily calculated from data used by heating and ventilating engineers that approximately 800 cubic feet of air at 60°F (15·6°C) would be warmed to 68°F (20·0°C) in taking up this quantity of heat from the body. The figure 68°F is used because it is the upper limit of the comfort zone for indoor temperatures, which has been shown to range from 60 to 68°F for our climate when ordinary clothing is worn.

"In this connection it is of interest to recall that a temperature of 65°F (18·3°C) has long been recognized as a desirable standard for the sick room; in fact it is on record that it was adopted one hundred and fifty years ago by Dr. William Withering when his failing health necessitated the control of the indoor climate of his house near Birmingham.

"Any gross departure from the percentage losses by radiation, convection or evaporation given above may lead to thermal discomfort. Thus, cold walls and uncurtained windows in winter time cause a sense of chilling due to excessive loss of radiant heat from that part of the body exposed to them. On the other hand, warm ceilings in top-floor flats and the radiation of heat from the inside surfaces of thin-walled hutments in summer may reduce heat loss by radiation to such an extent that conscious sweating is caused, unless convection losses are increased by fanning and additional ventilation."

THE NEED FOR WARMTH

To those of us whose dwellings are in temperate or cold climates the need for their warming in winter needs no emphasis. Crowden (1948) writes: "Due to sense of chill or excessive warmth, the distraction of thermal discomfort hinders mental work. We know from common experience that writing or any light work involving manual dexterity is not facilitated by cold hands. We are more susceptible to cold when doing such work because the little muscular effort needed only slightly raises our body heat production. Room temperature is therefore an important physical factor influencing the performance of such tasks; and it is worthwhile recalling that in an investigation of a light manual task in industry it was shown that in a room at 50°F (10°C) sedentary work involving manual dexterity took approximately 12% longer than when the room air had a temperature of 62°F (16·7°C). The air temperature in a house should not fall below 45°F (7·2°C) if condensation on walls is to be avoided; while the temperature in living rooms should be within the range of 60 to 68°F (15·6 to 20°C). In the kitchen, however, where more muscular work is done, a temperature of 60°F suffices. Our clothing habits to a large extent influence the air temperature levels deemed comfortable for the majority of people; but there is a wide variation between individuals in their subjective sensations of thermal comfort in the same environment."

Domestic Work

Sensations of fatigue are aggravated if the excess heat produced by muscular effort cannot easily be got rid of without conscious sweating. Air temperature, humidity and radiant heat are all concerned with body temperature regulation by the physiological processes which effect a balance between heat produced inside the body due to activity and the heat lost to the environment (Brown, 1958). That there should be some adjustment of the physical environment to suit varying states of bodily activity even in the home may be judged from the variations in body heat production in domestic work. Thus, in comparison with a body heat production of some 53 kcal/hour while sleeping, or 82 per hour while sitting at rest, a 9-stone woman may generate heat as follows according to the intensity and the nature of her muscular effort: knitting 95, ironing or dish washing 118, floor sweeping 138 kcal/hour. Ascending flights of stairs, energy is expended at much greater rates than these depending on the degree of haste. An obvious corollary from this is that for the aged, whose capacity for muscular effort is naturally less than that of younger persons, a minimum of stair climbing is desirable.

The Danger to Old Persons of Very Cold Housing

In an article on "Winter Ailments" in *The Practitioner* (Dec. 1964) Dr. Geoffrey Taylor refers to the "growing realization of the seriousness of the problem of hypothermia in the elderly", i.e. the physiological failure of old people to maintain body temperature within normal limits. This problem is examined in detail by Dr. Taylor who states that hypothermia in the elderly may be due to undernutrition and very cold conditions in their homes in winter. He notes that there are over one million elderly people in Britain living with financial aid from the National Assistance Board, and draws particular attention to the Ministry of Housing and Local Government Design Bulletin No. 1 on *Some Aspects of Designing for Old People*, 1962 (reprinted 1964), H.M.S.O. This bulletin merits careful study and action. The following extracts from it deal with some of the recommendations for the control of indoor climate in the interests of the elderly: "There are now five and a half million people over 65 years of age in England and Wales. . . . It is well known that old people, as they become less active, feel the cold more and need warmer conditions if they are to feel comfortable. They need warmth throughout the whole dwelling, and not only in the living room; to people whose movements have become slower with age, a warm bedroom and bathroom are especially important. Good heating eases many of the physical ailments, particularly rheumatism and arthritis, that trouble old people.

"Central heating is therefore generally accepted as desirable and has generally proved to be greatly appreciated by most old people. The heating system should be designed to maintain the temperature at a rather higher level than for family dwellings, say 70°F (21°C) rather than 65°F (18°C), and do so throughout the dwelling and, in grouped schemes, in the bathrooms as well. The distribution of heat is as important as the quantity. One aim should be to avoid chilling the feet, which restricts blood circulation and causes chilblains. A warm floor has advantages in this respect, though a floor temperature of more than 78°F (26°C) will be uncomfortable for the feet. Weatherstripping of doors and windows is also desirable to reduce floor draughts to a minimum. Variations in temperature should be kept down; within plus or minus 5°F (3°C) is acceptable. As the temperature of the body is higher at the head than at the feet the higher temperature should be provided at low level; this will also avoid a feeling of stuffiness."

PHYSIOLOGICAL EFFECTS OF DOMESTIC WORK UNDER HEAT STRESS

It is accepted that in hot weather, domestic tasks can be unwelcome and tiring. At the end of a day spent in performing the common, necessary tasks considerable fatigue is often experienced and the emotions arising therefrom may be freely expressed. Many investigations into the physiological effects brought about by working under heat stress have been made, and the effects measured by accepted instrumentation. But usually the severity of these test tasks performed by the human subjects has been so great, and the effects of the strains so unmistakable and so severe that such researches have been taken beyond the limits of discomfort likely to be encountered in the home. Brown *et al.* (1959) describe an investigation in which men and women carried out light physical work, comparable with that usual in kitchens, under different conditions of heat and moisture. It was possible to measure the small diminution of physical efficiency in these subjects after limited periods of only moderate exertion, and only moderate thermal discomfort. The subjects, four male and two female medical students, engaged in step climbing and in the several operations required in the washing of clothes. In the air-conditioned experimental room the dry-bulb temperature was gradually increased from 55°F (12·8°C) to approximately 100°F (37·8°C) and the wet-bulb temperatures ranged between 48°F (8·9°C) and 87°F (30·6°C). Before and again at the end of each work period the physical tone of the subjects was measured in terms of the Crampton Index. Crampton (1920) showed that a person's physical fitness, as affected for instance by an onset of reasonable fatigue, can be measured by taking the subject's systolic blood pressure and his pulse rate, after a period of recumbent rest before starting to perform the test task; and then repeating these measurements after the task has been completed. At the end of each set of measurements the Crampton Index is determined by means of a simple formula. It has been shown that a fall in this index indicates a drop in physical tone, or fitness. It should be noted that there is no "normal" Crampton Index for healthy persons, as there is a normal oral temperature. Individuals differ greatly in this respect; but the variations in the same person for Crampton Index are significant. Figure 3.1 shows records of some of the tests described above. The two top curves show the changes in Crampton Index of subject B who was washing clothes. The second set of curves show the summated sensations of heat, moisture and freshness as experienced during the variations of temperature and humidity recorded in the third pair of curves.† It should be noted that the solid line curves represent conditions whilst the amount of ventilation was controlled at will by the subject. The dotted line curves refer to an experiment when the ventilation was limited to the opening

---

† Brown and his co-authors give a table of the scales used in evaluating the sensations of heat and moisture as experienced by the subjects. Such summation was based on work by Bedford and also by Crowden and Lee (1940). The table has two columns, one for heat, and one for "moisture". The "heat" column shows +7 as unbearably hot, −7 as unbearably cold with zero as neutral. The "moisture" column shows +7 as unbearably moist, −7 as unbearably parched and zero as neutral. The ordinates in the second graph of Fig. 3.1 are derived from sensations enumerated on these scales.

of only one window. One particularly interesting conclusion given in this paper is that the correlation coefficients of transitory values of the Crampton Index both with the corrected effective temperature (C.E.T.) and the wet-bulb temperature are not significantly different. The C.E.T. is a function of the air temperature, the rate of air

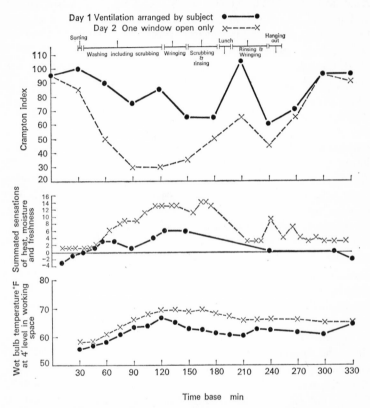

FIG. 3.1. Changes in the Crampton Index of circulatory efficiency and thermal sensations of housewife subject B in relation to variations in the wet bulb temperature during clothes washing.

movement, and the radiant heat gain or loss as well as the wet-bulb temperature. As has already been mentioned in Chapter 2, as temperatures rise, the wet-bulb temperature becomes a factor of greater importance than other influences on human thermal comfort.

Crowden (1948) has written: "It is necessary that the fundamental truths justifying the qualitative and quantitative definition of desirable standards for indoor climate should be clearly stated and under-stood. Such knowledge gives courage of conviction to those whose duty it is to insist that housing design, construction and equipment shall permit the ready attainment of desirable conditions by the occupants.

"Finally it must be emphasized that while the facilities for attaining desirable indoor conditions of warmth and fresh air ventilation are provided in modern houses, it rests with the occupants themselves to make proper use of them."

It may be said with little fear of contradiction on physiological grounds that cooling by the evaporation of water (as in sweating) is Nature's emergency cooling mechanism in man. A popularly understandable, but none the less impressive, demonstration of this was given in the publication, some few years ago, of an article in the American journal *Heating, Piping and Air Conditioning*. The ventilation by air conditioning of the dance hall of one of the junior prom. dances of the University of Wisconsin during the evening had been carefully controlled and the dry bulb and humidity of the air leaving the hall instrumentally graphed. The entering conditioned air was, of course, at an agreeable temperature. It was noticed that during each dance the humidity graph rose to a pronounced peak, which was followed during each interval by a fall to a value nearer to the entering humidity. There was a longer interval of lower humidity for the supper period; followed by another series of humidity peaks for each of the ensuing dances, with the expected long cooling off after the end of the entertainment. It is remarkable, when examining these graphs, that the rises which might have been expected in dry-bulb temperature for each dance were so small as to be hardly if at all noticeable. In these post-war days, work is not a very popular subject, but at least we may learn that "work" in a physiological sense may not necessarily be distasteful.

MEASUREMENTS OF GREAT AND OVERPOWERING THERMAL STRESS

Crowden (1949), as a result of experiences in the 1939–45 war, emphasized the importance of the control of working conditions for men in ships in the tropics. This applies also to men in armoured fighting vehicles. The classical experiments of George Fordyce and Charles Blagden, two medical men who lived for a very short experimental period in excessively hot rooms, together with Haldane's findings that the wet-bulb temperature is the arbiter of miners' capacity for work, provide evidence that the limits of men's ability to exist in excessive heat may be set by the ability to evaporate water. Crowden, who gives an extensive bibliography, quotes in particular

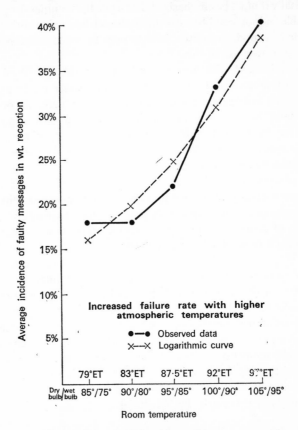

FIG. 3.2. Effects of hot and humid conditions on mental work (Mackworth).

the work of Mackworth (1946) who found that for young, fit and trained naval ratings, wireless operators, their ability to work efficiently was seriously impeded by too high an effective temperature.

He found that for men of this type there was a critical zone of effective temperature above which accuracy of performance of mental tasks declined (Fig. 3.2). This zone was from E.T. of 83°F (28·3°C) to

E.T. of 87·5°F (30·8°C). In one series of tests, in which the accuracy in wireless telegraphy was measured, it was shown that the average incidence of faulty messages rose from approximately 15% at 79°F (26·1°C) E.T. to over 30% when the effective temperature was 92°F (33·3°C). In view of the importance of restful sleep for the maintenance of day-to-day efficiency, other experiments were devised so that restlessness during sleep could be automatically recorded nightly for a period of a month. Observations made with six subjects (Fig. 3.3) showed that whereas they turned over heavily about

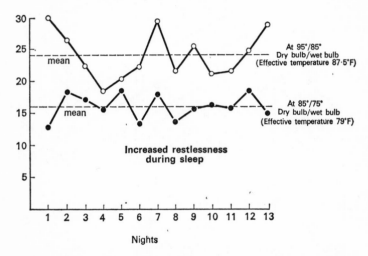

FIG. 3.3. Effects of hot and humid conditions on restlessness during sleep (Mackworth).

16 times during the night when the effective temperature was 79°F (26·1°C), the number of movements rose to 24 per night when the effective temperature was 87·5°F (30·8°C).

Limitations of Heat on Marching Soldiers

Eichna *et al.* (1947) performed experiments at Fort Knox, U.S.A. Thirteen enlisted men, white soldiers, all volunteers 18 to 30 years of age, marched round a track in a hot room for periods of 4 hours without rest at 3 miles per hour, wearing 20-lb packs. Physiological conditions and wet- and dry-bulb air temperatures were measured at hourly intervals. After some time complaints of headache, dizziness and nausea were common, but a man rarely dropped out during a 4-hour period.

Figure 3.4 records these results in a striking manner, showing the ambient conditions corresponding to three degrees of thermal stress and strain as evidenced in the subjects. The dry- and wet-bulb

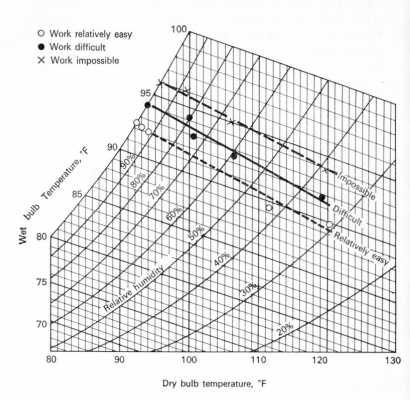

FIG. 3.4. Environments in which prolonged moderately hard work (300 cal/hr) was relatively easy, difficult, or impossible for acclimatized subjects. (From *J. Ind. Hyg. & Tox.* **27,** 1945, No. 3, 69.)

temperatures were entered on a hygrometric chart of the Carrier type: against abscissae of dry-bulb air temperatures are plotted ordinates of absolute humidity (not shown on this reduced chart). A series of lines sloping down from left to right indicates the wet-bulb temperature at any given air condition; in the same chart another set of curved lines, sloping down from right to left, gives the relative humidity at any point where dry-bulb and wet-bulb lines intersect.

A state of complete saturation corresponds to any point on the 100% relative humidity curve, where both wet and dry bulb show the same temperature. From the Fort Knox experiments the important fact emerges that for the three degrees of thermal stress, "relatively easy", "difficult", and "impossible", the corresponding air conditions when plotted lie close to the wet-bulb temperature lines 91°F (32·8°C), 94°F (34·4°C), and 95°F (35°C) respectively. These results are in close support of Haldane's classical observations with miners.

It may be noted that effective temperature is not considered to be an index of considerable thermal strain; it is much better adapted to measure the ordinary ranges of comfort and discomfort.

Safe Periods of Exposure to Extreme Environmental Temperatures

The *Guide and Data Book* of the American Society of Heating, Refrigerating and Air Conditioning Engineers (1966) gives a table showing safe periods of exposure for men, both when at rest and when working, under various temperatures and at various relative humidities.

Figure 3.5 gives a family of curves which show these times for safe exposures at various relative humidities and at a range of high temperatures. The physiological criteria used to establish these limits were based on the average time required to raise the pulse rate by 50 points from an initial rate of about 75 to 215; and the rectal temperature of 101°F (38·3°C) to 125°F (51·7°C) from initial levels of 98° to 99°F (36·6° to 37·2°C). It is stated that symptoms of distress do not appear until the pulse rate exceeds 130 and the rectal temperature 101°F (38·3°C) in exposure lasting more than 1 hour. It may be noted that the solid line in Fig. 3.5 represents the time of safe exposure for subjects working at the metabolic rate of 88 Btu/sq. ft of body area per hour.

STRESSES AND STRAIN OF EXCESSIVE HEAT

Leithead and Lind (1964) give a graph showing how a group of men became acclimatized to artificially heated conditions over a period of 9 days. They worked for 100 minutes each day at the rate of 300 kcal/hour. As the experiment proceeded the men found effort much easier, and there was a notable reduction in disagreeable sensation. After the third day there were considerable reductions both in sweat, pulse rates and rectal temperatures. The last two had been abnormally high during the first two days. On the other hand, by and after the third day the quantity of sweat lost had risen considerably. It is generally conceded by those who frequently go to the tropics for varying periods that fit men accustom themselves

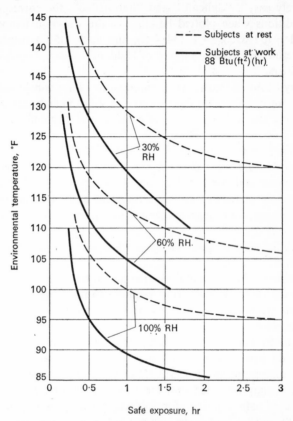

FIG. 3.5. Safe periods of exposure for extreme environmental temperatures. (Reprinted by permission from ASHRAE *Guide and Data Book*, 1965.)

in about 3 days to conditions of ordinary life. And it appears that one of Nature's compensations is that, for persons in normal health, the sweating mechanisms undertake this unusually hard task in about this period. In these days of rapid air travel, passengers to the East have none of the restful part acclimatization of a sea voyage; so newcomers should be gently used for a few days after arrival.

Measures of Heat Stress

It is accepted by physiologists that the amount of sweat produced is a measure of heat strain. McArdle *et al.* (1947) showed experimentally that it is possible to predict the amount of sweat that will be produced by a man under heat stress in a given time, from a knowledge of the dry- and wet-bulb air temperatures, the globe thermometer temperature, and the rate of air movement. From their results, McArdle and his colleagues devised the Predicted Four-hour Sweat Rate (P4SR) scale. They concluded that the E.T. scale was inaccurate as a measure of heat stress and that the sweat loss was the physiological measurement that correlated best with the severity of the experimental conditions. It is scarcely surprising, therefore, that it was considered worth while to construct a new scale in which the sweat loss was to be the index. Within wide ranges, the heat stress of any combination of dry- and wet-bulb temperatures, globe temperature, air movement, amount of clothing worn, and rate of work can be assessed from a nomogram in terms of the amount of sweat produced in 4 hours.

There is no doubt that sweat production is a good measure of the physiological strain experienced in response to a heat stress (Hatch, 1963). The Belding–Hatch Index is based on a comparison between the amount of sweat that has to be evaporated to maintain thermal equilibrium, and the maximum amount of sweat that it is possible to evaporate. The ratio of the evaporation required to the "maximum evaporative capacity" is the heat stress index for a given set of conditions.

The Wet-bulb Globe Temperature Index (Minard *et al.*, 1957) can be used to estimate thermal conditions likely to produce ill effects in man by undue stress. It was developed in the United States of America in order to evaluate a scale of heat stresses by which the training of recruits for the armed forces could be related to the tasks and the weather with which they were likely to be faced when sent to the tropics. It was said that as a result of this precaution many heat casualties were prevented, particularly among unacclimatized recruits for the Marines. The book by Doctors Leithead and Lind, cited above, gives a practical and useful guide to these new heat stress indices.

REFERENCES

BROWN, J. R. (1958) Physiological Reactions of Women to Heat and Humidity During Work in the Home. *Advanc. Sci.* **11**, 415.

BROWN, J. R., CROWDEN, G. P. and TAYLOR, P. F. (1959) Circulatory Responses to Change from Recumbent to Erect Posture as an Index of Heat Stress. *Ergonomics*, **2**, 262.

CRAMPTON, C. W. (1920) The Gravity Resisting Ability of the Circulation: Its Measurement and Significance (Blood Ptosis). *Amer. J. Med. Sci.* **160**, 721.

CROWDEN, G. P. (1948) The Physiological Aspects of Housing. *Health Education Journal*, **6**, 162.

CROWDEN, G. P. (1949) A Survey of Physiological Studies of Mental and Physical Work in Hot and Humid Environments. *Trans. Roy. Soc. Trop. Med. Hyg.* **42**, 325.

CROWDEN, G. P. and ANGUS, T. C. (1947) Modern Views in Ventilation, Warming and Cooling. *Practitioner*, **159**, 210.

CROWDEN, G. P. and LEE, W. Y. (1940) Sensations of Heat and Moisture. *Chin. J. Physiol.* **15**, 475.

EICHNA, L. W., ASHE, W. F., BEAN, W. B. and SHELLEY, W. B. (1947) The Upper Limits of Environmental Heat and Humidity tolerated by Acclimatized Men working in Hot Environments. *J. Indust. Hyg. & Tox.* **27**, 59.

Guide and Data Book (1965–6) Am. Soc. Heating, Refrigerating and Air Cond. Engrs., New York.

HATCH, T. (1963) Assessment of Heat Stress. In *Temperature, Its Measurement and Control in Science and Industry*. Reinhold, New York.

LEITHEAD, C. S. and LIND, A. R. (1964) *Heat Stress and Heat Disorders*. Cassell, London.

MCARDLE, B., DUNHAM, W., HOLLING, H. F., LADELL, W. S. S., SCOTT, J. W., THOMSON, M. L. and WEINER, J. S. (1947) *M.R.C. Royal Naval Report* 47, p. 391.

MACKWORTH, N. H. (1946) *Rep. Roy. Soc. Emp. Sci. Conf., June–July 1946*, vol. 1, p. 423.

MINARD, D., BELLING, H. S., and KINGSTON, J. R. (1957) The Prevention of Heat Casualties. *J. Am. Med. Assn.* **165**, 1813.

TAYLOR, GEOFFREY (1964) The Problem of Hypothermia in the Elderly. *The Practitioner* (Supplement: *Winter Ailments*), **193**, 761.

CHAPTER 4

DESIGNING FOR THE
WARMING OF BUILDINGS

SINCE the most important of all measurements for thermal comfort is the temperature as indicated by an ordinary thermometer, it is fairly obvious that when considering the warming of a building the most important data to be taken into account are the outside temperature and the desired inside temperature. So it is plain that the provision to be made for the heating of a house in a mild climate such as that of Cornwall would be unlikely to prove adequate for a similar house built in the north of Scotland.

ADJUSTMENT OF HEATING CAPACITY TO LOCAL CLIMATE

The differences in local climates as related to winter heating requirements are well documented. For this purpose climate classification is conveniently defined in "degree days" which have been determined for many localities. In this country, the outside air temperature that is taken as the basis for degree-day numeration is 60°F (15·6°C). During any 24-hour day throughout a year, or other period of time for which the degree days are to be enumerated, each degree Fahrenheit by which the outside mean temperature falls below 60°F goes to form one degree day. Thus for (say) a year any day on which the mean temperature has fallen to 50°F will add 10 degree days; whilst any 24-hour day on which the temperature has risen above 60°F will add nothing to the total. The year's degree day count is then the sum of all the average degrees Fahrenheit below 60° recorded for all 24-hour days. Figure 4.1, reproduced by permission of H.M.S.O. from a D.S.I.R. publication, was extracted by Dufton (1945) from the Meteorological Office's *Book of Normals*. It shows the degree-day distribution for parts of the British Isles. It is perhaps unexpected to find that equally severe winter conditions are likely over the small area on the top of Dartmoor as in the north of Scotland.

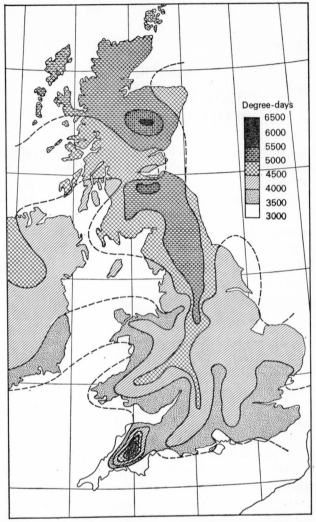

Fig. 4.1. British degree-days to a base temperature of 60°F (after Dufton). (Crown Copyright reserved.)

The number of degree days to be expected in any district can be used as a guide to the heating requirements of buildings in that district, but since the extremes of winter climate are of short duration, a system of heating load factors has been introduced. It is bad economics to install in any ordinary building a heating apparatus large enough to maintain the full desired indoor temperature for the whole of a heating season with lowest outside temperature. No one in ordinary good health is harmed by having a house a few degrees down for a few exceptionally cold days on which the full desired indoor temperature cannot be held. So whilst a heating plant should be able to deal with the usual winter conditions of any locality, and have a little extra capacity in reserve, it should not have to be unduly forced during a few exceptionally cold days; nor should it be needlessly large and costly for daily use.

Tables of degree days and load factors for various British districts, based on 10-year records from 1945 to 1955 are given in the I.H.V.E. Guide (1965). Heating "load factors" for any locality are the relative annual heat requirements as percentages of the heat that would be required in a year if the outside temperature remained at $30°F(-1 \cdot 1°C)$. These tables show that the degree days for Southampton are 3478 with 42% load factor; for Dundee the degree days are 4469, with load factor of 54%. So Dundee is not only colder than Southampton but the duration of the cold periods is longer. Additional useful information is provided in the I.H.V.E. Guide referred to above in a series of graphs giving, for a number of localities, the frequency of cold days and low temperatures recorded during past heating seasons.

With the readily available and recognized data, the calculations for heat losses from a building and of the necessary heat inputs are long but not difficult. Such work avoids any guess-work.

HEAT LEAKAGE FROM BUILDINGS

An extremely important consideration in the warming of buildings in cold weather, and one especially likely to be overlooked in our smaller dwellings, is the provision for retaining the heat that our various fires and radiators give out into the rooms. The heating of a building in cold weather has been likened to the pouring of water from a tap into a leaking bucket. To maintain a steady level of water in the bucket the water must be run in at a rate that makes up for the loss of water through the leaks. Here the level of the water to be kept in the bucket represents the elevation of the indoor temperature above that of the outside. The flow of water from the tap into the bucket is the input of heat from fires, or other heat sources. The greater the rate at which heat is put in, the warmer will be the house. On the other hand the water leaks represent the loss of heat through walls, roofs, windows and the ground floors; to this must be added the heat unavoidably lost with the ventilation air. Sir Alfred Egerton said at the Conference of the Institute of Fuel (May 1956): "The design of buildings with a view to reducing the heat loss is, I consider, of equal or even of greater importance than actual efficiency of heating appliances. . . . In this country there are millions of detached and semi-detached houses of very low standards of insulation and leaky to air and heat. . . . The extra cost of insulating in new housing is negligible compared with the value of advantage gained over the years—in fact in a very few years."

A striking example of the usefulness of heat-loss calculations occurred during the 1939–45 war. A large ground-floor factory, in use for the manufacture of important parts for electronic instruments, was heated by over-head gas-fired radiant heaters, whose rated heat output should have been sufficient to secure winter comfort. But the building became too cold for the efficiency of the operatives. It was found that the very extensive roof was composed of unlined asbestos-cement sheeting, and calculation showed that the heat losses through such a roof were excessive. It was planned to install extra boilers, pipes and radiators, for additional central heating; but this would have entailed a considerable interference with production, and other inconveniences, especially serious in those days. Using tabulated data, calculation showed that lining the roof with a proprietory cellular insulating sheeting would enable the existing gas heaters to keep the factory comfortably warm. This lining was fitted with little inconvenience; and the situation was saved.

Legal Requirements for Thermal Insulation

Regulations for the insulation of industrial buildings whose erection was begun on or after 1st January 1959 are contained in the Thermal Insulation (Industrial Buildings) Act, 1959. The Act, which applies to all parts of Great Britain other than Northern Ireland, prescribes that for standard roof insulation the thermal transmittance coefficient, or U-value, shall not exceed 0·3 Btu per sq. ft. per °F per hour. There is a special provision if the inside design temperature is lower than 70°F.

The Value of Wall Insulation in Houses

From the above it is evident that the importance of control of heat losses through roofs can be of major importance. But in

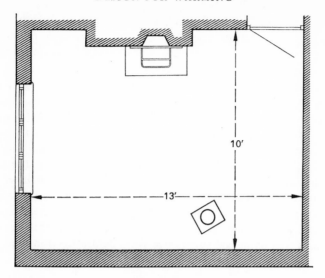

Plan of room

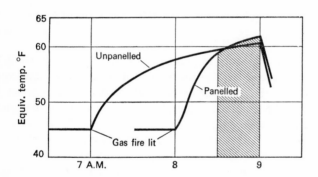

Fig. 4.2. Room used for half an hour. (Reproduced by permission of the proprietors of *The Philosophical Magazine*.)

dwellings, both old and new the losses of heat through walls can also be serious. Dufton (1931) gives an example of how the losses of heat through the brick walls of an ordinary residence were reduced by lining these walls with oak panels. The measure of this heat saving was given by the rapidity with which the room could be warmed after the panelling was fitted.

Figure 4.2 shows the plan of the room and also two curves which indicate how much more rapidly the room could be warmed, after it had been panelled, by the same gas fire as previously used. After panelling, the equivalent temperature rose from 45°F (7·2°C) to 52°F (11·1°C) in half the time that was needed with the plain walls. Note that the equivalent temperature is the criterion here.

The obvious advantage of such insulation is the saving of heat losses and fuel, but there is another, and one less obvious. With a plain brick and plaster wall in an unheated room the inner surfaces of the walls are cold, and this, apart from the cold air, will have a cooling effect on the occupants and their clothing. If the walls are heat-insulated, much of the radiant heat from a radiant fire is re-radiated back into the room and its contents because the inner surfaces have gained heat which warms them. Such returned heat affects human comfort, and is indicated by the globe thermometer and so recorded as an increase in equivalent temperature. This work of Dufton's was published twenty-five years before Egerton made the statement quoted on page 42.

PREDETERMINING THE HEAT LOSSES

Heat Losses through Walls, etc.

The design of heating installations is based on the amount of heat, generally expressed in British thermal units per hour, that will be lost from the building, as designed and finished, after it has been warmed to the required inside temperature and exposed to the anticipated winter weather. These rates of heat loss, the thermal transmittance coefficients, or "U"-values as they are termed, are tabulated in the I.H.V.E. Guide (1965), expressed in British thermal units lost per hour from a wall, or other boundary of a building, per sq. ft of area per °F of temperature difference between the proposed indoor and outdoor temperatures. The prepared tables give not only the "U"-values for many structures and materials used for walls, roofs and windows, but also for their degree of exposure to weather and sun, and their orientation. If a wall faces north-east the heat loss is greater than that of a similar structure facing west. Such finer points in calculation are of importance when large buildings are being designed. For example, the rate of heat loss through window glass is higher than through the usual wall; but it is shown that by using double glazing, with air space, the heat loss is about halved.

FIG. 5.3. Radiant fire, glass enclosed; gives convected heat to room
and hot water to house.

FIG. 5.4. Gas radiant fire, with convection air warming.

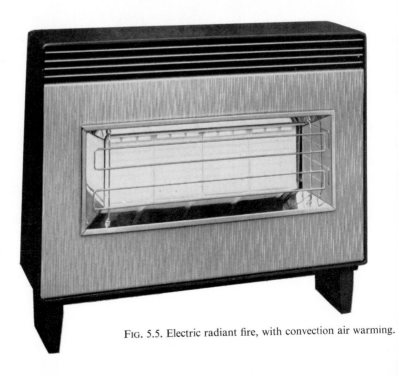

FIG. 5.5. Electric radiant fire, with convection air warming.

FIG. 5.6. Gas fire with simulated log blaze and regained convection.

FIG. 5.7. Turbo-type air heater: warmed air projected through louvres.

Heat Losses through Ventilation

In the consideration of the warming of buildings it is true to say that the two questions of heating and ventilation can never be entirely divorced. A very important factor in the heat losses from buildings is the loss of heat in the warmed air and its replacement by cold air from outside. Such ventilation, when fortuitous, is termed "infiltration". The requirements for ventilation are dealt with in Chapter 6; and in Chapter 5, in which various methods and mechanisms for heating are discussed, ventilation, both the planned and the fortuitous, is considered. It will be shown that with many of the older solid fuel domestic fires one of the greatest faults found is excessive ventilation caused by the venting of too much of the warmed air of the room with the products of combustion. The outside air that replaces this must be warmed unless discomfort is to be tolerated.

Heat Lost in Vented Warm Air: Example. Consider a smallish living room 12 feet by 10 feet by 8 feet high. Cubic content = 960 cubic feet. Assume that conditions in the room are comfortable when the dry-bulb is 65°F (18·3°C) and the relative humidity 40% which corresponds to a wet-bulb of 52°F (11·1°C). From tables, the volume of one pound of dry air is 13·57 cubic feet in these conditions, so the weight of the dry air in the room is 960/13·57 or 71 lb; the weight of the contained water vapour is only about 0·005 lb. The total heat of such air, which includes the latent heat of evaporation of the contained moisture is given as 13·71 Btu per lb of dry air.

Assume also that the condition of winter outside air is D.B. 32°F (0°C) with relative humidity 92%. Its total heat will be 3·78 Btu per lb of dry air. And if this air replaces the warm air of the room once per hour the hourly heat requirement will be 71 (13·71−3·78), or 9·93 × 71 = 705 Btu per hour. Or, if the heating is by electricity, nearly three-quarters of a unit per hour.

DESIGNS FOR POST-WAR HOUSES AS INFLUENCED BY HEATING AND VENTILATION REQUIREMENTS

The need for more houses and the inevitable difficulties in their provision called for a reconsideration of the building by-laws of local councils, to find whether economies were possible by reducing the dimensions of the new houses to be built and by using different materials. There were strong opinions held that a reduction of ceiling height in rooms from 8 feet to $7\frac{1}{2}$ feet would be prejudicial to health. On the suggestion of J. M. Mackintosh, formerly Dean of the London School of Hygiene and Tropical Medicine, a series of experiments were carried out at the Ministry of Works Test Unit to determine whether physiological or psychological ill-effects could be noted and recorded in the occupants of rooms of heights less than the accepted 8 feet. Crowden (1951) describes these researches. Simultaneous and identical measurements were made in two experimental rooms, one of which was 8 feet high, and the other made so that its ceiling could be lowered to $7\frac{1}{2}$ or to 7 feet. Crowden's paper presents charts which show that in respect of heating both by electric and coal fires, with windows shut, and in winter, the final temperatures in both the 8-foot and the 7-foot rooms were the same within 2 or 3°F ($1 \cdot 1$ or $1 \cdot 7$°C). The same was found to be true for equivalent temperature, effective temperature, and mean temperature of surrounds: all were well within the accepted comfort zones.

During the same series of experiments records of the rates of ventilation, as measured by air change, were made in the two rooms. Graphs of these ventilation rates are plotted (Renbourne *et al.*, 1949) for various methods of heating, and with different openings of chimney throats and windows. These graphs show that the air changes for the two rooms were generally the same within half a change per hour. These air changes ranged from $4 \cdot 7$ to $3 \cdot 8$ in the 7- and 8-foot rooms respectively. To quote Crowden: "It was shown that there was no material difference in air change under identical conditions of heating and ventilation provision, while in either room fresh air ventilation conforming to accepted standards for health could readily be obtained by making appropriate use of windows according to the number of occupants."

In addition to these experiments a number of field enquiries were made in houses with rooms 7 ft 6 in. high in Hertfordshire, Sussex, Herefordshire and Denbighshire. Crowden concludes: "The general findings of the field survey tend to confirm the experimental results. It is of interest to note that most of the subjects used in the experiments and tenants of the houses were quite unconscious of the fact that the rooms were lower than usual. While tall individuals, or householders possessing large furniture and pictures may dislike rooms that are lower than 8 feet it may be safely assumed that a height of 7 ft 6 in. is generally satisfactory."

REFERENCES

CROWDEN, G. P. (1951) The Height of Rooms in Dwellings in Relation to Health and Comfort. *J. Roy. San. Inst.* **71**, 108.

DUFTON, A. F. (1945) Degree Day Chart. *Post-War Building Studies, No. 19, Heating and Ventilation of Buildings.* H.M.S.O.

DUFTON, A. F. (1931) The Warming of a Room. *Phil. Mag.* **11**, 1233.

I.H.V.E. Guide (1965) Institution of Heating and Ventilating Engineers, London.

RENBOURNE, E. T., ANGUS, T. C., ELLISON, J. McK. and JONES, M. D. Assisted by Croton, L. M. (1949) The Measurement of Domestic Ventilation. *J. Hyg.* **47**, 1.

CHAPTER 5

THE NATURE OF HEAT FOR COMFORT, AND ITS PRODUCTION

THERE is no doubt that the brightness of the solid fuel fire and the variability of its flickerings and of the heat that it gives out are generally found to be pleasing. It may be asked if these effects are solely due to psychological causes. But even if this were so it would be a bold person in these days who would say that, for this reason, these attractions of a bright fire are of no account. It is physiologically true that the steady and continuous excitation of any nervous receptor will lessen that receptor's sensitivity. If prolonged, such steady stimulation can result in monotony and fatigue, so that in time the sensation becomes blunted. Examples have been found in ventilation systems which have produced continuous air flow over persons in one direction at very low velocity. Effects of this kind have been guarded against in the design for the ventilation of the rebuilt House of Commons. Here warmed and conditioned ventilation air is delivered across the floor of the House from a series of jets which are situated behind the members, and well above their heads. It is the members seated on the side of the House opposite to the jets who directly benefit from their action. This is to direct a gentle current of tempered air over their faces and the upper parts of their bodies. The small unstable eddies of low velocity so produced are known to give sensations of freshness (Bedford, 1944). Further to prevent monotony, at automatically controlled intervals, dampers switch the air delivery from the jets on the one side of the House into the jets on the opposite side. In this way there is a periodic alteration in the degree of air movement. Neither the members of the Government, nor those of the Opposition are favoured at each other's expense.

ADVANTAGES OF VARIABLE STIMULI

It is known that the nervous mechanisms of the body appreciate changes that are not violent. According to Hill (1920): "The coal fire is a good ventilator, venting 18,000 to 20,000 cubic feet of air per hour: it is variable—not monotonous—and cheerful to look at, and for a given output of heat the first cost of coal is only about 40% of gas (Bone). The gas fire is not such a good ventilator as an open fire, and is monotonous. The attention a coal fire requires provides some measure of change and exercise to a sedentary person." It should be noted that the construction and performances of both solid fuel and gas fires have been considerably improved since the above was written. But unfortunately Leonard Hill's proposal, made a long time ago, that gas fires should be provided with some device that would cause them to emit heat, and perhaps light, with some irregularity was not acted upon at the time (but see page 54).

PRACTICES IN DOMESTIC WARMING

When considering the theory and practices of domestic heating it is not possible to separate these from the needs of ventilation; though in a sense ventilation and warming may be said to be in opposition. For when air is changed in a comfortably warm room, in cold weather, much valuable heat may be vented with the vitiated air (see also Chapter 4). Bedford and Warner (1939) give the requirements for "a pleasant and stimulating atmosphere", as indicated by a statistical examination of many observations made by himself and a colleague. Such requirements are:

(1) A room should be as cool as is compatible with comfort.
(2) There should be adequate air movement. At the room temperatures customarily maintained in winter in Great Britain the velocity should be about 30 feet per minute; velocities below 20 feet per minute tend to cause feelings of stuffiness. In summer weather, or in hot factories, higher rates of air movement than those mentioned are desirable.
(3) The air movement should be variable rather than uniform and monotonous, for the body is stimulated by ceaseless change in the environment. Ventilation by open windows is likely to give variable air movement, but in mechanical ventilation installations the inlets should be so designed, and the velocity of discharge so arranged, that suitable eddy currents are set up.
(4) The relative humidity of the air should not exceed 70% and should preferably be much below that figure.

(5) The average temperature of the walls and other solid surroundings should not be appreciably lower than that of the air, and should preferably be warmer. The combination of cold walls and warm air often causes feelings of stuffiness.

(6) The air at head level should not be distinctly warmer than that near the floor, and the heads of the occupants should not be exposed to excessive radiant heat.

(7) The air should be free from objectionable odours.

THE TRADITIONAL OPEN FIRE AND ITS MODERN MODIFICATIONS

In spite of many modern tendencies the liking for the open fireplace burning soft coal, coke or smokeless fuel, dies hard in Great Britain. And this is especially true in the industrial North, where cooking is often carried out in a general living room, and where outside temperatures in winter are commonly lower than in the South. In the older houses in the North a solid-fuel cooking range is an essential part of the living room, or kitchen-living room, as it has been termed. G. K. Chesterton has called our open fire "The veritable flame of England, that is still kept burning in the midst of a mean civilization of stoves."

Faults of Open Fireplaces and Simple Improvements

As already stated considerable losses of valuable heat take place if a dwelling has excessive ventilation.

In Chapter 6 are described a number of experiments which were made with the object of deciding important questions of design for new residences then urgently needed. A small part of these experiments described by Angus (1949) was concerned with the measurement of air change in experimental living rooms, as influenced by fireplaces, and the air that enters rooms fortuitously through building faults, and through ventilating devices.

It was found that improvements could be effected by a number of simple measures, such as the use of better-fitting doors, windows, and proprietary weather stripping under doors. Hopper openings in the upper parts of windows, with alterable aperture give good renewal of air, and, with a curtain inside, prevent undesirable draughts after dark. An alternative is to make use of constant-flow air bricks. Amongst other facts was found that with current building practice there can be considerable air leakage around doors and windows where their frames are joined to the brickwork of the walls. The usual space between the bottom of a door and the floor is equivalent to the free area given by a standard ventilation air-brick. Special methods were employed to measure the rate of air flow from a room into the fireplace, with the fire alight, and also to evaluate the relative door and window crack resistances to the passage of air. It was found that there is advantage to be gained by drawing air for the fire from the roof-space through vertical ducts. But it is doubtful whether the elaboration of building involved in the making of such ducts will be considered worth while.

These experiments were done with open fireplaces of both old and new types. A fireplace such as that shown in Fig. 5.1, a type that has been put into new houses until quite recently, was found to have many causes of inefficiency. The large chimney, with only a small constriction at the throat, caused excessive ventilation of the room. The necessary combustion air, with the smoke, was accompanied by additional air from the room, which was replaced by air which had to find its way into the room through fortuitous openings and imperfect joins in the building joints; much of this air entered through the space below the door, which in practice can rarely be less than $\frac{1}{4}$ inch wide. As much as 6,000 cubic feet of air per hour was measured passing from one of the model rooms into the chimney. By accepted standards, four persons require only 2,400 cubic feet of outside air per hour; so in this instance an extra 3,600 cubic feet per hour of presumably warmed air was passed up the chimney. The heat so lost is considerable.

Modern Fires

The type of fireplace shown in Fig. 5.2, known as free-standing, not only passes much of the otherwise wasted heat back into the room in the form of air warmed by convection, but also, having no high metal frame in front of the fire itself, sends more radiant heat into the lower parts of the room. The smaller chimney throat with adjustable shutter prevents excessive ventilation. Provision is made for allowing the necessary air to enter without causing draughts.

Figure 5.3 shows a modern development of the once-open fire; here the fire is screened by a window of special heat-resisting glass, transparent to radiant heat. The supply of combustion air, with the heat output of the fire, are controlled by means of a damper

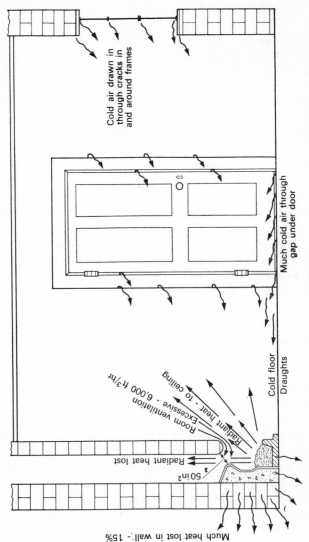

Cold air drawn in through cracks in and around frames

Much cold air through gap under door

Cold floor

Draughts

Room ventilation — 6,000 ft³/hr
Excessive heat — to ceiling
Radiant heat lost
50 in²
Radiant heat lost

Much heat lost in wall — 15%

Fig. 5.1. Living room—winter—ineffective fire. Heat lost up chimney and uncontrolled draughts.

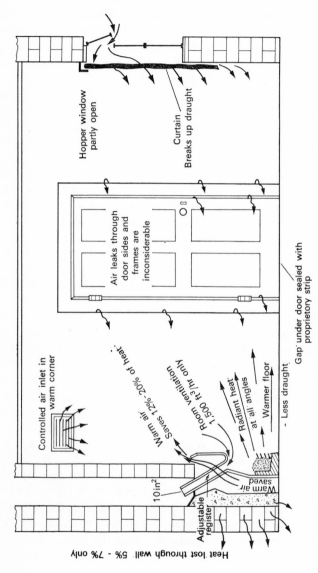

Hopper window partly open

Curtain Breaks up draught

Air leaks through door sides and frames are inconsiderable

Gap under door sealed with proprietory strip

Controlled air inlet in warm corner

Warm air Saves 12%–20% of heat

Room 1,500 ft³/hr ventilation only

Radiant heat at all angles

Warmer floor

– Less draught

10 in²

Warm air saved

Adjustable register

Heat lost through wall 5% - 7% only

Fɪɢ. 5.2. Living room—winter—freestanding fire. No excessive ventilation, draughts reduced.

adjustment. The lower side openings through which extra air enters and the upper ones through which it is recirculated after warming can be seen in the photograph. Some of these fireplaces are provided with back-boilers in which water is heated for domestic use. Some manufacturers state that by this means several radiators about the house may be supplied with hot water, giving a partial central heating. A typical English compromise!

Figure 5.4 shows a modern radiant gas fire of attractive design. The products of combustion are vented into a chimney of the usual type; but before they are so disposed of as much heat as possible is taken from these gases and returned to the room as warm air through the top grille.

Figure 5.5 shows a radiant electric fire of appearance in keeping with modern trends. In this fire too, any heat finding its way back from the hot elements is regained in convection air. One advantage of this arrangement is that, being emitted in front the hot air is unlikely to stain walls.

Figure 5.6 shows another popular and efficient type of gas fire. In this one not only is there provision for the return to the room of convected heat, recovered from the products of combustion and the hot metal surfaces, but the occupants of the room are not denied the satisfaction of seeing a "live" burning bright fire. Below the steady-burning gas radiants is an imitation "fire" of timber logs and white ash of burnt wood. But the attractive coloured flames and the red and yellow flickering glow is produced by mains-operated electric glow lamps under the dummy logs. The light from these lamps is caused to vary, and the colours to change, by means of shutters and colour filters mounted on rotating frames which are moved by small windmills operated by the heat from the glow lamps. Not so very long ago the rivalry between the gas and the electric industries was so keen that on no account would any electrical device be embodied in any gas appliance. But for this attractive "gimmick", as for other purposes with a common value, there has been formed an alliance. It has been stated that when the innovation of the dummy fire was first suggested to the gas authorities there was opposition in some quarters. But when the gas-electric log fire was introduced into the market it was immediately taken up by the purchasing public. Electric radiant fires are also made with imitation dummy coal or log fires in front of the heat-giving glowing elements.

DOMESTIC HEATING BY FORCED-CONVECTED AIR

When air is heated or cooled by being mechanically driven over hot or cold surfaces it is said to be subjected to "forced convection". The air heated by being drawn through the so-called radiators of automobiles is so heated. The application of this well-known principle to small scale domestic heating is new.

Figure 5.7 shows a typical turbo-type air heater. Electrically heated resistance coils heat air that is drawn rapidly through them, and then ejected at high velocity generally in a flat, wide stream along the floor through a louvered opening. Some of these heaters may be regulated to give different outputs of heat, even as the usual electric radiant fires may be set to use one or more "bars". This type of heater does not suffer as much as many other sources of hot air from the tendency of heated air to rise into the higher levels, where generally it is of little value. It is a fact that an air jet blowing along a flat surface such as a floor tends to carry farther than a jet blown into a free space; this, with the momentum of the moving air, helps to carry the hot air from this heater to where it is likely to most appreciated, at foot level.

Such small heaters are useful for warming chilly corners of large apartments. A novel feature of the unit is the design of its centrifugal fan. This does not revolve in a close-fitting shroud; it also has an elongated impeller with blades that somewhat resemble the familiar cutting blades of a lawn mower. This design gives a low blade-tip velocity, with quiet performance.

BACKGROUND WARMING WITH RADIANT HEAT

Fishenden and Willgress (1925) described researches carried out for the Gas Industry to help in the development of gas fires. The emission of radiant heat from fires tends to be unidirectional and so there are difficulties in keeping a large room comfortably warm by this means. In these researches the subjects who gave their subjective sensations of heat and cold sat in front of gas fires of variable heat output.

The temperature of the ambient air of the room was varied and the subjects gave their opinions as to when the resultant of the air and radiant temperatures gave three ranges of thermal sensation: "slightly too cool", "comfortable", and "slightly too warm".

Figure 5.8 gives a chart derived from this research. The intensity of radiant heat required to give these three degrees of thermal sensation with air temperatures ranging from 45°F (7·2°C) to 68°F (20°C) can be determined. From such data the makers of radiant fires should be able to determine the intensity of radiant heat required to give comfort in any part of a house warmed to a specified temperature by some system of background warming.

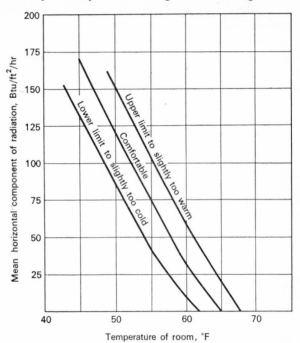

FIG. 5.8. Radiation required for comfort with different air temperatures.
(Crown Copyright reserved.)

That many people show a real liking for a small, local additional source of radiant heat, even when the general room temperature has been raised to accepted comfort levels, is shown by the results of inquiries recorded by Welch (1960).

CENTRAL HEATING

There is no question that in Great Britain small-scale domestic central heating has come to stay, and that it will become increasingly popular as time passes. For a great many years central heating has been used exclusively in public buildings, schools, large offices, hospitals and the like. But the acceptance of central heating for private houses is comparatively new. The wastefulness of the solid-fuel open fire weighs against its extended use. In addition to this, the virtual extinction of the domestic servant makes the tasks of laying and lighting solid fuel fires, with the inevitable handling of solid fuel and ash, increasingly unwelcome. These facts, added to the undesirability of smoke both in the house and in the outside air, have enhanced the trend towards domestic central heating. Both the Gas Council and the oil companies now specialize in designing and directing the installation of central-heating systems for smaller buildings. Such heating is most convenient, being clean and entirely automatic in action and control.

Central Heating by Electricity

Central heating by electricity is also gaining popularity. In recent years a great concession has been made to consumers of electricity, in that they are encouraged to take as much power as possible during the "off-peak" periods, when prices for electricity are much lower than the prices for current used at other times. Electric clocks, with switches, are supplied with the meters that measure and cost the loads. Heat from off-peak electricity can be stored in hot-water tanks well insulated; and the hot water is circulated through the house radiators under thermostatic control, as the heat is required. In the same way, the smaller quantities of hot water needed for baths and other domestic uses can be heated over-night and, at a small extra cost, can be given an extra temperature "boost" at off-peak rate between 13 and 16 h., a great convenience. Another aid to central heating by electricity is the use of "block heaters", in which off-peak current heats large masses of building brick or fire-brick contained in neat sheet-metal cases. These hot bodies can be placed in halls, or in rooms that become too cool by night or day. The slow emission of low-grade heat from such storage supplies a grateful warmth. There can be little doubt that the increasing use of atomic power for the generation of electricity will encourage the use of off-peak current. We are informed that, at present, it is not economic to run atomic generators at low load; so these new power stations are likely to be run all night at full load, the conventional stations being brought into service as the daily loads of factories, railways and other users are called for.

Hot-air Central Heating

A widely used method of central heating for houses in North America, especially in early days, employed a large wood-fired or coal-fired basement furnace, surrounded by a sheet-iron casing. Air heated in this casing was distributed by thermal circulation to parts of the house as desired. Cool air through grilles at floor level was ducted back to the basement furnace, and so reheated. A very agreeable dry heat was thus obtained. Needless to say the whole of the house was well heat-insulated all over, double glazing following at a later date. Air humidification was sometimes employed, for if the outside air was very cold, it would have a very low dew point: then when heated to over 70°F it might well be excessively dry, and wooden furniture and fittings often suffered. In Great Britain a somewhat similar hot-air heating system is now coming into use for new dwellings. But instead of the basement furnace there is usually a solid-fuel, oil-fired, or gas-fired heat exchanger through which air is made to circulate by means of a quiet centrifugal fan. This arrangement is becoming popular.

Individual Central Heating in Blocks of Flats

An ingenious device called the SE-Duct flue system has been developed which enables the tenants of separate flats in large multi-storey blocks to have and to control their own central heating. The fuel used is town gas, which is metered to each flat. The heat, which includes that for the hot water needed for domestic purposes, is produced in gas-fired heating appliances, which, although situated inside the flat, are completely sealed off from the air in the rooms. Instead of drawing the necessary combustion air from within the room, and venting the products of combustion through separate chimneys, in the SE-Duct system each gas-fired heating appliance takes its air from, and vents its burnt gases into, the same large vertical duct, a number of which pass through the entire block of flats, from below basement floor to simple terminals on the flat roof. For the combustion-air supply to the gas flame no opening into the occupied living spaces is needed, nor is such permissible. The air inlets to the gas-burning heaters open into the sides of the large vertical ducts. Experience has shown that, with correct design, combustion in the heaters on the upper floors is not adversely affected by dilution of their combustion air with burnt products arising from the appliances in the lower storeys. The needed outside air is provided through horizontal air ducts which run below the basement floor, and into which the large vertical ducts connect, and from which they draw the necessary unburnt air. Each exterior end of these horizontal ducts extends beyond the outside wall of the building, where it opens to the air through some suitable opening.

Figure 5.9 gives a good general illustration of the principle, with an indication of the manner in which the heating appliances are arranged in the flat interiors. The central heating may be carried out by hot-water circulation through the usual radiators, or by air heating heat-transfer apparatus, fired by sealed-off gas burners, the rooms being heated by the circulation of heated air. One great advantage of the system is the large saving of horizontal inside area due to the absence of separate chimneys. Another considerable advantage in modern town-planning is the absence of smoke.

Welch (1960) describes how the SE-Duct flue system was developed for unusually large blocks of multi-storey flats in the New Town of Harlow, Essex. The designs were due to the Gas Council Research Centre, in consultation with the Harlow Development Corporation, architects, and builders. Welch's paper contains many other items of information (see pp. 15 and 16) that are likely to be of interest to anyone concerned with domestic gas services. He quotes the indoor temperatures chosen for many flats, and for other gas-heated residences, in which the occupiers could control the temperature by means of thermostats. He also notes that many persons like to use small radiant electric fires in addition to their central heating by gas. Since its inception the SE-Duct system has been adopted, not only in other British towns, but also in France and in Australia.

Floor Heating

In recent years the space heating of buildings by means of heat stored in solid floors and emitted by radiation and convection from the upper surface has come into use. Such floors can be heated by embedded hot pipes, or electric heaters. Off-peak electric heating has been applied here, but difficulties may be encountered, because the high temperature of the mass required to store sufficient heat to make good all losses during day and night, may raise the surface temperature of the floor undesirably. This can cause discomfort, and damage to floor coverings (Barton, 1963).

Central Heating: the Distribution of Heat, and Stratification

In the majority of central-heating systems, other than those employing warmed floors or ceilings, the room or other space is heated by convection; and the enclosed air, heated by hot pipes or radiators, generally rises up the walls and diffuses throughout the room. In this way the whole room, its boundaries and its furniture are warmed to a pleasant temperature. But convection heating has one important disadvantage: it is most difficult to guard against stratification, because the heated air ascends to the ceiling, and tends to stay there. Then, as the room warms up, the heat works downwards, and the lower parts of the bodies of the occupants are cooler than their heads—a condition that has often been shown to be undesirable. Thus Bedford and Warner (1939) write: "Such methods of warming, which give distinctly greater warmth about the head than near the floor, are undesirable, for, besides causing the feet to be chilly, such a condition produces distinct feelings of stuffiness. These effects are produced by any local chilling of the feet, whether it be due to temperature gradients or to draughts." Members of Parliament frequently complained about the ventilation of the House of Commons. That chamber (1939) was ventilated by the propulsion of air through the perforated floor and its subsequent extraction through the ceiling; and Hill (1920) ascribes the sensations which were experienced there to the rapid passage of air currents over feet and legs: "In our field investigations in factories we have frequently encountered systems of heating which have caused the air at head level to be 6–8°F (3·4–4·5°C) warmer than that near the floor, while on some occasion we have observed differences as great as 10 or 11°F (5·6 or 6·2°C). In such workrooms we have regularly experienced feelings of stuffiness and have found the conditions distinctly unpleasant. Similar undesirable effects are caused if the head is exposed to excessive radiant heat."

An excellent way of minimizing stratification is to install radiators on the wall below windows. Room air cooled by contact with the glass of the panes, as well as cold outside air leaking in through the inevitable cracks around the window frames, will produce descending currents at the windows. Unpleasant floor draughts can therefore result; but if this descending air is met by a current of warm air from a radiator, a more or less horizontal current of warm air, some of it likely to be from outside and so adding to

FIG. 5.9. Large block of flats, with individual central heating and hot water, by the SE-Duct system.

Fig. 5.10. Central heating by convection. Gilled tubes, concealed behind cover; air enters under, comes out of top slit.

the natural ventilation of the room, is distributed in a desirable manner.

Ceiling Heating

Much work has been done in this country on the space heating of interiors by hot panels with surfaces flush with the ceilings, or with the whole ceiling warmed to a suitable temperature by embedded series of hot pipes or electric heaters. From what has been written above of the ill effects of hot air stratification, it might well be thought that heating from warmed ceilings would suffer seriously on this account. But this is not so. What occurs in practice is that there is a thin layer of quite warm air, only a very few inches in depth, in contact with the ceiling; and that nearly all the heat diffused into the room and on to its occupants is in the form of low-temperature direct radiation from the hot or warm ceiling surfaces. This radiation warms the floor and the furniture, and helps to warm the air by heat convected from these warm surfaces. The effect is most agreeable.

Certain precautions must be taken when designing ceiling heating systems. It is most important to be sure that adequate heat insulation is provided in the upper layers of the heated ceiling. Failure here results in a great loss of heat in the top storey rooms; in intermediate floors, poor insulation of the ceiling can cause uncomfortable heating of the floor of the room above. An instance has been cited of an uncomfortably hot floor in a hospital ward, on account of unwanted heat finding its way through the ceiling of the rooms below.

Another important consideration is concerned with the discomfort caused by too much intensity of radiation on the head. This is a possibility to be borne in mind when all the known heat losses from an enclosure are to be met by the heat input from a heated ceiling. Bruce (1953) and Chrenko (1953) have dealt fully with this problem and the calculations by which undue heating of the head may be avoided. Particular difficulties may be encountered in low-ceilinged rooms.

Concealed Central Heating

With both ceiling and floor heating, the exposed surfaces that supply the heat to the enclosure are essential parts of the room and so should be quite unobtrusive. In most of the conventional central heating systems the radiators are usually rather large objects, and so difficult to harmonize with domestic decorations. Fairly recently methods of heating by convectors situated around the wainscot of the walls of rooms have been introduced.

Figure 5.10 illustrates a room fitted with a type of wainscot heating. This is unobtrusive and takes up very little space. Behind a narrow covering with slit opening at top and bottom there are situated horizontal hot pipes, very often carrying gills. Air passes up through these gills, is heated by convection and then ascends as a warming influence over the walls and windows. It is stated that such installations are particularly efficient in the distribution of heat.

REFERENCES

ANGUS, T. C. (1949) The Study of Air Flow, Ventilation and Air Movement in Small Rooms, as affected by Open Fireplaces and Ventilating Ducts. *J. Inst. Heat. & Vent. Engrs.* **17,** 378.

BARTON, J. J. (1963) *Electric Space Heating.* Geo. Newnes, London.

BEDFORD, T. (1944) Heating and Ventilation: Requirements and Methods. *Brit. J. Indust. Med.* **1,** 31.

BEDFORD, T. and WARNER, C. G. (1939) Subjective Impressions of Freshness in Relation to Environmental Conditions. *J. Hyg.* **39,** 498.

BRUCE, H. H. (1953) Panel Heating, Some Practical Applications. *J. Inst. Heat. & Vent. Engrs.* **21,** 507.

BRUCE, H. H. and CHRENKO, F. A. (1953) Heated Ceilings and Comfort. *J. Inst. Heat. & Vent. Engrs.* **20,** 375.

FISHENDEN, MARGARET and WILLGRESS, R. E. (1925) *The Heating of Rooms.* Fuel Research Board Tech. Paper No. 12. H.M.S.O., London.

HILL, L. (1919, 1920) The Science of Ventilation and Open Air Treatment. *M.R.C. Special Reports,* No. 32, 1919, and No. 52, 1920. H.M.S.O., London.

WELCH, W. H. (1960) Gas, Coke and Water Space Heating Services in Houses and Flats; Harlow New Town. Lecture to the Institution of Gas Engineers, 14th Nov. 1960.

CHAPTER 6

VENTILATION

THE purpose of ventilation is the removal of vitiated air, and its replacement with fresh air. For long it was assumed that the air of occupied enclosures became polluted, in particular by carbon dioxide. It was also held that expired air contains some subtle poison of organic origin, and of uncertain effects. This theory was conclusively proved to be untrue by Hill (1920). The concentration of carbon dioxide in country air is in the neighbourhood of $0 \cdot 03\%$, and is sometimes a little higher in the outside air of a town. In a really "stuffy" and crowded room $0 \cdot 1\%$ of CO_2 may be measured. In fact this concentration was found, on a winter's evening, of all places, in the refectory of a post-graduate school for doctors! Such a concentration is harmless in itself; but it is usually accompanied by disagreeable heat, moisture, and odours, so that before long the occupants take some action. It is recorded that sailors in submerged submarines experience no distress until the CO_2 concentration reaches about $3\frac{1}{2}\%$. During experiments on the natural, chimney-produced ventilation of living rooms (Renbourne et al., 1949), in which carbon dioxide was injected into the experimental room to concentrations up to 4%, an experimenter experienced no breathlessness during the short periods for which such concentrations were maintained.

But, expecting unusual symptoms of some kind, the experimenter realized, on consideration, that he was breathing unusually deeply whilst making and recording observations: a very light form of physical effort. There was no after-effect. Such is Nature's unconscious compensation for meeting minor changes in the chemical composition of air. But it should be noted that during this experiment, unlike the men in a submarine, the experimenter was not exposed to heat, moisture or odour.

The following is an extract from Hill's account of his classical experiment:

"A wooden chamber was constructed for me at the London Hospital Medical School with large glass observation windows and a man-hole for entering. Inside were electric heaters, coils through which cold water could be circulated, and three electric fans for stirring the air. Seven or eight students were shut within until the CO_2 reached 2–4%, and the oxygen fell to 17–16%, and the wet-bulb rose to 80–85°F ($26 \cdot 7$–$29 \cdot 4$°C), and the dry-bulb a degree or two higher. The students went in laughing and chatting, but as the wet-bulb rose became flushed and moist and ceased to talk. One might try in vain to light a cigarette, and borrow matches from another, unaware that the percentage of oxygen had fallen too low for combustion. The breathing was deepened but no headache resulted from short exposure to 3–4% CO_2. Their discomfort was greatly relieved by putting on the electric fans and whirling the air in the chamber. They asked for the fans to be put on when they were stopped. The air enmeshed in their clothes was saturated at body temperature; whirling the air of the chamber of 85° wet-bulb, through the clothes and over the skin of the face, relieved the heat stagnation. The fans lowered the cheek temperature of one student from 34° to $31 \cdot 5$°C. The pulse rate frequency of seven students was lowered by the fans thus: 92 to 72; 96 to 86; 128 to 84; 94 to 74; 106 to 100; 72 to 52; 92 to 86. Breathing the air of the chamber through a tube by one outside caused no discomfort. . . . When the CO_2 was put into the chamber so as to suddenly raise the percentage to 2, the subjects were unaware of this."

PRESENT-DAY STANDARDS

Many would be surprised to learn that it is on no delicate chemical or physical measurement that our standards of air-change ventilation are based; but upon the sensitivity of the human nose to odour. On determinations made by this homely instrument are based the quantities of outside air that should be supplied to each individual in an enclosure per unit of time. The experimental work by which these standards (which also apply in this country) were established was carried out in the United States of America (Yaglou *et al.*, 1936). One of the experimental methods was the enclosure of groups of persons of different socio-economic status in a metal-lined room, in which the ventilation rate could be carefully controlled. Referees, straight from breathing the outside air, came into the test room and gave their assessments of odour intensity in accordance with an agreed scale.

This scale, which is shown in Table 6.1, is the scale on which it was found that agreement could be reached among a team of trained referees. In another series of experiments (Lehmberg *et al.*, 1935) a man was enclosed, lying down, in a large box painted white internally and carefully cleansed. Measured volumes of air were supplied at one end of the box, whilst at the other end referees, often ignorant of the nature and the purpose of the experiment, placed their noses at the open end of a pipe through which the air left the box. These persons gave their estimations of the odour intensity. As a result of these series of experiments an interesting and important fact was established: The less crowded a room the less the volume of outside

TABLE 6.1. Scale of Occupancy Odour, as agreed by a number of trained judges (Reprinted by permission from ASHRAE *Guide and Data Book* (1965).)

Odour intensity index	Characteristic term	Qualification
0	None	No perceptible odour
$\frac{1}{2}$	Threshold	Very faint, barely detectable by trained judges; usually imperceptible to untrained persons
1	Definite	Readily detectable by all normal persons but not objectionable
2	Moderate	Neither pleasant nor disagreeable. Little or no objection. Allowable limit in rooms
3	Strong	Objectionable. Air regarded with disfavour
4	Very strong	Forcible, disagreeable
5	Overpowering	Nauseating

air per minute required to keep down odour intensity. So in this respect the popular idea that a large room needs less ventilation than a small room is vindicated.

A possible reason for this was found by some of the same teams of workers in further experiments. It was proved that the aerosols which compose the sources of body-odour are very unstable; so these odours fade rapidly with time, and in as short a time as 2 minutes there is a considerable diminution in their intensity.

Now it has been shown that in a room at ordinary temperature there is above every person an ascending current of air that has been warmed by his body. If then the air surrounding a person rises at once to ceiling level in a large room, it may have time to lose much of its odour before fortuitous circulation returns it down to breathing level. The present-day recommendations, as accepted in the U.S.A., for minimum rates of air change sufficient to avoid unpleasantly high occupancy odour level are quoted in Table 6.2.

TABLE 6.2

Type of occupants	Occupancy density, cubic feet of room space per person	Outside air supply, cubic feet per minute per person
Heating season, with or without recirculation, air not conditioned		
Sedentary adults of average socio-economic status	100	25
	200	16
	300	12
	500	7
Labourers	200	23
Grade school children of average socio-economic status	200	38
Children attending private grade schools	100	22

Heating season. Air humidified by means of centrifugal humidifier. Water atomization rate 8 to 10 g.p.h. Spray water changed daily. Total air circulation 30 c.f.m. per person.

Sedentary adults	200	12

Summer season. Air cooled and humidified by means of spray dehumidifier. Spray water changed daily. Total air circulation 30 c.f.m. per person.

Sedentary adults	200	< 4

RECOMMENDED STANDARDS OF VENTILATION FOR BRITISH DWELLINGS

Table 6.3, which is taken from an official Report on the Heating and Ventilation of Buildings (1946), gives the recommended British standards for dwellings.

The Report from which Table 6.3 has been extracted gives the following comment on the subject of house ventilation: "The rate of ventilation is commonly expressed by giving a figure for the number of complete air-changes per hour in each room, assuming by one air-change that a quantity of air has entered and been withdrawn equal to the total volume of the air in the room. To estimate the number of air-changes required for any one room, it is necessary to assume the number of persons using the room and the quantity of air required per person. The exact minimum amount of ventilation required for health cannot be accurately determined, but after considering the available evidence on the subject the Committee has adopted a figure of 600 cu ft per hour per person as the normal minimum requirements for the purpose of this Report."

Figure 6.1 shows two graphs giving the rates of air change per person recommended for use in this country by the Institution of Heating and Ventilating Engineers.

In Chapter 5 some attention has been directed to present-day problems of heating and ventilation for dwellings and what are hoped to be some practical suggestions have been given.

TABLE 6.3. Recommended Minimum Standards of Ventilation in Dwellings

Room	Air supply per hour
Living room (four persons)	2400 cu. ft
Bedroom (two persons)	1200 cu. ft
Bedroom (one person)	600 cu. ft
Kitchen (cooking for not more than six persons)	one air change
Halls and passages	one air change
Bathroom and W.C.s	two air changes (or more)

METHODS OF VENTILATING DWELLINGS

Plans for new buildings, perhaps sometimes a little grandiose, are widespread, and it is important that these plans should provide for adequate ventilation of inhabited rooms, properly under the control of the occupants. As has already been shown there is evidence that excessive ventilation is inherent in certain features of modern house construction, and that adequate ventilation can be obtained with ceilings only 7 feet above floor level instead of the more usual 8 feet (see Chapter 4). But however great the demand for new houses it should be remembered that there are still in use great numbers of old houses in Great Britain, easily convertible to present-day ideas of health and comfort.

The Restoration of Old Houses

A great many houses erected before the 1914–18 war were very well built with regard to what architects term "the body of the house". They are still standing and lived in. In some respects, and in regard to amenities such as heating, hot water and even sanitation they leave much to be desired, and some had been condemned as uninhabitable by Local Authorities. In the City of Leeds in 1955 no less than 22,000 old houses were to have been demolished, having been declared unfit. In addition 20,000 back-to-back houses had

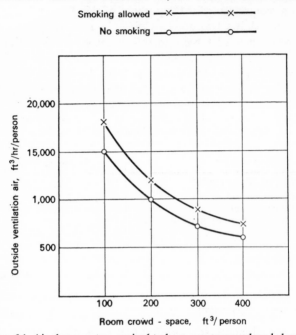

Fig. 6.1. Air change rates required to keep occupancy odour below an objectionable level, at various crowd densities.
From figures given in the *Guide* of the Institution of Heating and Ventilating Engineers, 1965, London. By permission.

been condemned. One of the shortcomings of this type of house is the difficulty of securing adequate ventilation, and it was on these grounds that many of these dwellings had been condemned. However the Leeds Council carried out an investigation of the ventilation rates in these houses, under the advice of Professor G. P. Crowden of the London School of Hygiene and Tropical Medicine. As a result of this work, many of these old houses which were of otherwise sound construction were brought up to standard and saved.

The following extracts from the 1955 Report of the Leeds Medical Officer of Health, Dr. I. G. Davies, are applicable to many old buildings.

FIG. 6.2. Automatic air-brick natural ventilator. Silent moving vanes prevent excessive ventilation in strong winds.

FIG. 6.2. Automatic air-brick natural ventilator. Silent moving vanes prevent excessive ventilation in strong winds.

Inlet Ventilation. "The infiltration of air between doors or windows and their frames may be considerable, especially when they are ill-fitting. However it is undesirable to rely upon fortuitous leakage from these sources to obtain the necessary inlet ventilation. Adequate inlet ventilation has been secured by provision of a 9 × 9 inch or 9 × 6 inch air-brick in the external wall of each room.

"It has been found that in times of high wind an uncontrolled air-brick may cause a draught, and is, therefore, liable to be papered over. For this reason a constant-flow natural ventilator has been fitted to the inside of the air-brick opening. It should be emphasized that this constant flow ventilator is not an air impeller, or mechanical ventilator, and its only object is to control and direct *natural* air movement."

Outlet Ventilation. "As regards *outlet ventilation,* living rooms and large bedrooms with an existing fireplace and flue may be adequately ventilated by this means as an outlet ventilator. In the case of kitchens, where the recommended minimum rate of fresh air supply is relatively high, an outlet is necessary, and in these cases a disused set-pot flue has been adapted to act as outlet ventilator.† This set-pot flue is a most efficient extractor if connected with a ventilator set in the kitchen ceiling, and extended above the roof with an anti-down draught chimney pot. In the case of small bedrooms which have been converted into bathrooms, the provision of one 9 × 9 inch inlet ventilator alone is not sufficient to produce the necessary two air changes per hour; some outlet ventilation is also necessary. Where the new bathroom is at the corner of the block, the position enables a second ventilator to be placed in the gable wall or side wall of the bathroom, so that in the bathroom there is a diagonal cross ventilation. Where the new bathroom is not at the corner of the block,

---

† In most of the smaller houses, as in cottages, especially in those built some 80 years ago, the only means of obtaining any considerable quantity of hot water, for the washing of clothes and for other purposes, was provided by a large bowl-shaped basin, of copper or of iron, which was built into the kitchen or scullery in a brick enclosure containing a coal-fired grate under the basin. This coal fire had to be provided with a flue, and a chimney opening above the roof. With modern heating developments these coal-fired water heaters—termed "coppers" in the Southern Counties, "set-pots" in the North—fell into disuse. But experience has shown that their chimney flues can be adapted to provide excellent extract ventilation in kitchens, where the substitution of gas and electric cooking for solid fuel, with the absence of any flued fire, may result in inadequate ventilation.

the inlet ventilator in the face wall can be assisted by an outlet ventilator through the ceiling into the roof space."

Making use of Staircase Ventilation. "The ventilation of the staircase helps the ventilation of all the rooms in a back-to-back house. If the staircase happens to be against a gable wall, the fixing of an air-brick right at the top of the staircase helps the staircase to act as a gentle extractor from all the rooms. Where the staircase opens into an attic, in which there is an existing fireplace and flue, there is an even better effect. When the attic has no fireplace and flue, and is to be used as a bedroom, then it should be improved by the provision of a dormer window, with a constant-flow ventilator in the cheek of the dormer."

The Measurement of Rate of Ventilation. The method of measuring the rate of ventilation of rooms used in the investigation of the back-to-back houses was that developed by the Department of Applied Physiology of the London School of Hygiene and Tropical Medicine. This easy and convenient method is known as the Carbon Dioxide Direct Analysis Technique. Briefly, carbon dioxide, suitably warmed, and dispersed into the air of the room under test to a concentration of up to 4% has its percentage in the air measured at 15-minute intervals. By application of a logarithmic formula the rate of air change can be calculated (Renbourne *et al.*, 1949).

The following is an extract from Dr. Davies' report on the results obtained by this method.

"It is most interesting to compare the measured rates of ventilation obtained in an improved Type II back-to-back house with those of a modern semi-detached house. The ventilation rate in the scullery of the type II back-to-back house after improvements as described was 2·14 air changes per hour. That measured on the same day in the kitchenette of a modern semi-detached house built in accordance with the building by-laws was 0·94 changes per hour. The air changes in the living rooms of the same two houses were measured on 20th July, 1955. In the improved back-to-back house it was 1·59 air changes per hour; and in the modern house only 1·06."

Natural Ventilation in Ordinary Houses

Carne (1946) described a long investigation of the natural ventilation found in the upper rooms of ordinary houses in London. Each room had an open fireplace, and the usual sash windows which were kept closed during the tests. Wall openings of the air-brick type were made in the wall on the same side as the window. Air change was measured by the rate of decay of carbon dioxide, which was injected into the room until the initial concentration was about 3%. Chimney flue openings were varied in area from zero to 50 sq. in., and the velocity and direction of outside winds carefully measured. It was found that the air change depended almost entirely on the wind; effects of inside-outside temperature difference were negligible. In no experiment was a fire alight below the chimney. The rooms were considered to be closed and unheated; in fact much the same as many bedrooms as used in this country. There is one important deduction to be found here. Modern houses are frequently built without flues and chimneys communicating with the open air; a tendency likely to grow with central heating. As Carne showed, any room with an open fire grate is quite well ventilated, even if only one window is partly open, or if there is an air-brick in the wall. In many modern houses air-bricks are required in outer walls, but here a disadvantage exists that if a strong wind blows on the side of the house so ventilated such an unpleasant cooling will occur that the occupants will cover the opening. Carne gives an analysis of wind effects on and around an exposed building, and where side winds can exert an aspirating action on a chimney. He also shows that a wind on the side of a house that is pierced by an air-brick can cause a considerable influx of air, even if the wind direction is far from normal to the wall.

Bedford *et al* (1943) made investigations on lines similar to those of Carne. They found that in flats built in 1936, where all the rooms examined were of less than 1000 cu. ft capacity 78% of the observations showed ventilation rates of less than one air change per hour; 15% less than 0·4 change per hour. At the other end of the scale, in cottages built about 1860, much higher ventilation rates were found; the same is true for older houses. A substantial amount of leakage may occur around windows which are apparently close-fitting. A room in a university building, with well-fitting steel-frame windows, had a rate of air change with door and windows closed of 0·36 per hour, but with the cracks round the windows and door sealed with glued paper the ventilation rate was reduced to 0·08 air change per hour. Bedford and his colleagues investigated the effects of air-bricks, with and without the forces of winds, and they also confirmed the large ventilation value of open flues with no fires under.

INDUSTRIAL AND SPECIAL VENTILATION

Factories

The ventilation requirements specified in the Factories Act of 1937 are of a general character, and intended to secure that every workroom shall be adequately ventilated. Few numerical requirements are given, but, as in other applications, much is left to the experience and the good sense of H.M. Inspectors of Factories; and it may be added that these officers are usually found by managers to be helpfully constructive when suggesting needed improvements. The general provisions of the Act are contained in Section 4, which requires that:

"All workrooms must be adequately ventilated, and that sufficient fresh air must be circulated for the purpose. Numerical standards are not laid down, and up to the present there are no legal decisions which might assist heating and ventilating engineers to determine the capacity of heating and ventilating plants necessary for complying with the law. In the heating season, if sufficient air movement is provided to be refreshing without creating draughts, fresh air is generally required only in amounts sufficient to prevent palpable body-odour, provided that there is no contamination by dust, fume or other impurity. Such amounts depend upon the number of the occupants and upon the degree of their activity. For light sedentary occupations, even with the maximum density permitted by the Factories Act, a comparatively small quantity—about 550 cu ft per person per hour—has been found sufficient. This quantity is regarded as an absolute minimum, and it is desirable to design for a net rate of not less than 750 cu ft of fresh air per person per hour."

The *Guide to Current Practice* of the I.H.V.E. (from which the above notes have been extracted) gives many more particulars, and references to special requirements, where high humidities, dusts and fume have to be considered.

Exhaust Ventilation

It is seldom realized how limited is the suction effect of a current of air entering an exhaust fan, or a hood connected thereto. When air comes out of an orifice in the form of a jet, this jet may have a long carry, and hold its velocity through a considerable distance. It is very much otherwise when air is being drawn into an opening. In fact, in popular image, unless the hand is placed nearly into an exhaust opening it may be hard to know whether there is any air movement. The velocity of a stream of air entering an exhaust opening falls off inversely as the square of the axial distance from the centre of the opening. This fall in velocity has been compared with the fall in electrical potential at increasing distances from a charged body. Data are available for hoods and other non-circular exhaust openings. Dalla Valle (1952) describes some of the fundamental researches on these problems in which Hatch collaborated, and much of the fruit of their work, and that of later investigators, is available in the I.H.V.E. *Guide*, and in its American fellow, the *Guide and Data Book* already mentioned. An important parameter is what is known as the capture velocity needed to entrain various solid particles when air-borne. To exhaust a dust particle in an air stream, the stream's speed must be sufficient to snatch the particle away before it has time to fall, and before it may be swept aside by chance side draughts. Thus in factories where exhaust systems are in operation it may be important to assure sufficient entry of outside air to feed them.

Hospital Operating Theatres

Angus (1959) attempted to summarize the requirements of surgeons in this important matter, and quoted changes in the temperatures required in operating theatres that have taken place during the last 30 years or so (see p. 76).

REFERENCES

Angus, T. C. (1959) Air Conditions in Operating Theatres. Part 1. *The Med. Press*, **242**, 113.
Bedford, T., Warner, C. G. and Chrenko, F. A. (1943) Observations on the Natural Ventilation of Dwellings. *J. Roy. Inst. Brit. Architects*, **51**, 7.
Carne, J. B. (1946) The Natural Ventilation of Unheated "Closed" Rooms. *J. Hyg.* **44**, 314.
Dalla Valle, J. M. (1952) *Exhaust Hoods*. The Industrial Press. New York. 2nd ed.
Hill, L. (1920) The Science of Ventilation and Open Air Treatment. *M.R.C. Special Report* No. 52. H.M.S.O., London.

LEHMBERG, W. H., BRANDT, A. D. and MORSE, K. (1935) A Laboratory Study of Minimum Ventilation Requirements: Ventilation Box Experiments. *Heat, Piping and Air Cond.* 7, 44.

RENBOURNE, E. T., ANGUS, T. C., ELLISON, J. McK., and JONES, M. S. (assisted by Croton, L. M.) (1949) The Measurement of Domestic Ventilation. *J. Hyg.* 47, 1.

YAGLOU, C. P., RILEY, E. C. and COGGINS, D. I. (1936) Ventilation Requirements. *Trans. Amer. Soc. Heat. & Vent. Engrs.* 42, 133.

CHAPTER 7

THE EFFECTS OF EXCESSIVE INDOOR HEAT IN TEMPERATE CLIMATES, AND ITS CONTROL

THE control of cold and damp in dwellings is comparatively simple and inexpensive in lands where fuels are readily available. But the mitigation of great heat and drought, and the control of indoor climate in lands of great heat, with or without excessive humidity, is a much more difficult problem. It may be said that really effective solutions have been found only in recent times.

It is accepted that life for Europeans in many parts of India and Africa is trying, and for those whose working life has been spent in such lands, long periods of leave have been considered to be necessary to offset a general deterioration in health and energy. Excessive heat, usually lasting for only a few days during the summer, is commonly experienced in many temperate climates. Markham (1947) terms such limited periods of excessive heat "heat handicaps"; and adds that for London, the heat handicap amounts to at least three days per year: days on which the air temperature is over 76°F (21·1°C). It is only in maritime areas that temperature extremes between summer and winter are least marked. Markham adds that warmer countries, such as France, Italy, Greece and Mexico will have much greater heat handicaps. Also that so far no method has been devised for the cheap effective control of hot weather. "It will perhaps be generally admitted that at present methods of cooling and dehumidifying a house to any extent are beyond the means of most workers in any country. Air conditioning, fans, and dehumidifiers are, it is true, becoming more and more common as a feature of the houses of the well-to-do in the U.S.A. and Canada, but even this is not sufficient to justify us in considering the control of high temperatures and their concomitant of high radiation as within the powers of the masses of any country."

In Great Britain artificial cooling and refrigeration become necessary during periods of high temperature for the preservation of food, and also, probably, for the operating theatres of hospitals. The very high temperatures of 80°F (27°C) that not very long ago were considered by the medical profession to be necessary for patients undergoing operations are now seldom required. In one London hospital the surgeons were dissatisfied if the operating theatre temperature rose above 70°F (21·1°C). There is always a considerable quantity of what is termed "wild heat" entering the air-stream of any forced ventilating system; so to cool the necessary clean ventilation air for operating theatres, when the outside air temperature is above the required level, calls for some refrigeration (Angus, 1959).

MITIGATION OF OVERHEATING IN FACTORIES AND OTHER BUILDINGS

Insolation

In excessively hot weather, which in England sometimes occurs in May and June, though not often for more than a few consecutive days, considerable relief may be obtained by whitening roofs. This remedy is simple and inexpensive; the whitening is easily renewable in localities where smoke and the like quickly soil surfaces. There is a recipe for a durable whitewash such as is particularly useful on roofs of galvanized iron or asbestos-cement sheeting. It is seldom realized that these two cheap and useful building materials have almost identical, and very low, heat insulating properties—equally important in cold weather and when under hot sun. Unless effectively whitened, roofs made of these materials transmit much solar heat. Crowden gives interesting figures of the effectiveness of whitewash on corrugated iron roofs exposed to West African sun. His notes and observations are given in the Appendix, pages 101–3, with practical directions and recipes for whitewashing.

Unless some such precautions are taken these thin roofs, especially if unlined, act as large radiators on persons below them.

Present-day buildings, particularly schools and tall office blocks, often have walls mainly composed of glass; the heat from the sun can be excessive in such buildings. The well-known "greenhouse" effect is due to the fact that glass is transparent to light radiations and to the short infra-red, in which most of the sun's heat reaches us, but not to radiation of longer wavelength. On entering a glazed building the solar radiations are immediately absorbed on solid surfaces which they encounter, and heat them. The heat thus gained is partly passed to the enclosure by convection of the air, and partly re-radiated at a much longer wavelength, as dark heat. All this dark heat is retained in the enclosure, some being taken up by the occupants, their clothing and other internal objects and the rest absorbed by the window glass, which is opaque to it.

Reflecting Blinds

Some years ago, at the London School of Hygiene and Tropical Medicine, difficulty was experienced from overheating in upper-floor lecture rooms in the afternoon. These rooms had large windows facing west into which the sun shone above buildings on the other side of the road. A remedy was found in providing, inside the windows, roller blinds of fabric, the outer, or window sides, of which were lined with bright aluminium foil. Unfortunately no temperature observations were made at the time, but the improvement was unmistakable. The theory of this method is that the solar radiations having passed in through the glass are reflected out again, not having suffered change of wavelength. It is questionable whether blinds of a matt surface and pure white would be equally effective; they might be, but a practical difficulty would be that they would soon become soiled, whereas aluminium as now manufactured keeps its polish very well. Here is another practicable device, which, despite an inexpensive and promising pilot trial has not been further developed. Another modern development which is widely used is found in large Venetian blinds, composed of light thin strips of plastic material. It is likely that if the outer surfaces of such blinds were suitably whitened or highly polished they would form excellent protection against solar heat for modern buildings. Being Venetian blinds they do not interfere with window ventilation.

But it has been pointed out that for the protection of the interiors of buildings from the heat due to insolation by means of slat-type blinds, the angle at which these slats are exposed to the sun is critical to their thermal performance (Markus, 1964). For various reasons it appears to be impracticable to provide exterior protection for the windows of any but small buildings in Great Britain. Sun breaks, such as are described in the next chapter for use in tropical houses, would, for one thing, prevent adequate indoor lighting in gloomy weather. The only outside protection likely to be used here is the familiar whitewashing, or colour washing, of the glasshouses used in horticulture.

Control of Solar Heat through Glass Windows

Ordinary commercial glass is transparent to long-wave ultra-violet radiation, visible light, and the short-wave infra-red up to about 2·5 microns. These radiations predominate in the insolation received at the ground. But, as already stated, glass is very opaque to the infra-red radiations longer than these. There are some proprietary heat-resisting glasses now available (Window Design and Solar Heat Gain, *B.R.S. Digest*, 1966). One grade of heat-absorbing glass is of a dark green colour and absorbs infra-red heat, but only transmits about 50% of the visible light. This is useful for roof lights, and where light glare must be reduced, and if colour does not matter. Another grade of glass has about 75% of light transmission with only a slight greenish tinge, but transmits over 20% of solar heat. It should be remembered that about 30% of the absorbed heat can be transmitted into the building from the heated inner surface of such a window pane. Unless protected, a pane of this kind acts as a vertical radiator panel, not only radiating long-wave infra-red heat into the room, but also warming the inside air that is in contact with the glass. A practical alleviation is obtained by using double-glazed windows, preferably with the outer pane composed of heat-resisting glass. If then the enclosed air space connects with the outside by means of suitable small top and bottom openings in the window frames a chimney effect results; much of the air heated by the sun-warmed outer pane now finds its way out at the top before transmitting its heat to the room via the inner pane.

Heat-reflecting (rather than heat-absorbing) glasses are in course of development. If successful, such glasses should not become much hotter than the outside air, even when exposed to the sun. Another advantage of double-glazed windows, as already mentioned, is that they retain heat in winter time. The question of the ventilation provided by opening windows to prevent undue heating in summer is discussed in the *B.R.S. Digest* (1966) referred to above.

PERSONAL COOLING IN WORKPLACES BY AUGMENTED AIR MOVEMENT

Air movement over the person has an important influence on thermal comfort. Hill's kata-thermometer played a large part in early investigations of these effects. This instrument is influenced by dry-bulb air temperature and the velocity of the air past the bulb, but it is now known that the effect of air movements on its readings may be misleading in their relation to their effects on human comfort because its bulb is so much smaller than the human body. Nevertheless there is evidence that changes of air velocity between 10 and 30 ft/min can have significant effects on human comfort; and in the early days the kata-thermometer was considered to be an instrument of prime importance in the measurement of thermal conditions. So highly was it regarded that after 1919 at least two large industrial concerns, each with large factories in Great Britain, regulated the internal conditions in these factories by kata-thermometer standards. Moreover, the originators of the Effective Temperature Scale (see Chapter 2) enjoined the use of the kata-thermometer for the determination of the air-velocity factor.

In using the kata-thermometer the physical quantity measured is the cooling power of the environment on the kata bulb in millicalories per square centimetre per second; it is usually denoted by H. Hill, in papers quoted in previous chapters, had shown that for indoor comfort at sedentary work a value of $H = 6$ satisfied most people, and that the "grumble point", or the state at which rising temperature combined with increasing air stagnation caused feelings of oppression, occurred at about $H = 4$. For lower values of H, increasing heat stagnation brought increasing distress.

Cooling Marine Engineers

The importance of air movement was illustrated by investigations carried out on board a large Atlantic liner (Angus, 1936) in which the engineers on the engine-room control platform were exposed to considerable heat from the steam machinery. To improve conditions the Company fitted large uncased centrifugal fans providing 1·3 changes per minute of air drawn in from the outside and distributed at high speed and in irregular currents over the engineers. Angus took readings of the dry-bulb temperature and the kata cooling power H at a number of positions on the control platform, in the stokehold of the oil-fired boilers, and in the dynamo room. Some of his results, together with the deduced values of the air velocity, are shown in Table 7.1. During these observations the temperature of the outside air drawn in by the engine-room fans was 55°F (12·8°C).

The large fans on the control platform were much appreciated. Not only was the temperature lowered when they were started, but the dry kata reading was also considerably reduced. The stokehold was ventilated by a plenum system, separate from that of the engine room. It was agreed that both the engine room (with fans on) and the stokehold were comfortable by steamship standards even in hot weather. It will be noted that the kata factor H was well above the value of 6 given by Hill for comfort in sedentary conditions, and thus apparently good for moderate activity.

Thermal Discomfort due to Draughts

The dynamo room was considered to be the most uncomfortable part of the engineers' tweendecks. This is of considerable interest since the observations show that the temperature at shoulder level

TABLE 7.1

Position	Temperature °F	Dry kata cooling factor H	Air velocity ft/min
Engine-room control platform			
Fans stopped	80·5	3·9	55
Fans running	75·0	7·3	180
Stokehold (oil-firing)			
Shoulder level	71·5	7·9	150
Dynamo room			
Head level	77	4·9	66
1 ft above deck	77	7·3	220

was only 2°F above that in the engine room with fans running whilst the conditions near foot level were almost identical with those in the engine room. But, as the figures show, the air velocity at head level in the dynamo room was much less than at foot level, with the consequently much lower cooling power ($H = 4·9$ as against $7·3$). It is to this fact that the discomfort in the dynamo room is attributable, providing a good example of the condition, always to be avoided, of hot head and cold feet.

NATURAL VENTILATION IN ENGLISH FACTORIES:
AN APPROXIMATE MEASUREMENT
AND THE EFFECT OF CHANGES

Another investigation that showed the importance of adequate air movement was carried out in a large factory making tyres for trucks and autos (Angus, 1936). The tyres were moulded in auto-claves: a very hot process. Two moulding shops were in use during the same period of hot summer weather. An interesting comparison was made possible because the only difference in conditions was due to the amount of natural ventilation provided in each shop. An approximate measure of the magnitude of natural ventilation was arrived at by dividing the volume of the building, in cubic feet, by one thousand times the total free area in square feet provided by all such openings as windows, doors and roof vents. This simple, and not very precise measure, termed the "Opening Figure", was found, in this instance, to give figures of manageable sizes, with meaningful results, for comparing the reactions of the occupants.

The older moulding shop had an opening figure of 2·32 sq. ft per thousand cubic feet contents. The newer shop, of later design, no doubt influenced by previous experience, was provided with much larger side openings, giving, together with the roof openings a total opening figure of 8·6. In the older shop, during the period July to September, the dry kata-thermometer cooling power only once rose to near 6, and was frequently below 2; and the men certainly suffered by reason of excessive heat. But in the newly-built shop, where precisely the same process was carried out, the cooling power never dropped below 4, and was usually above 5.

PERSONAL FANNING

The benefits accruing to operatives in hot industries from mechanically induced air movement have been recognized, and as will be described below the same applies to residents in the tropics. Angus (1936) describes a device known as the jet fan (Fig. 7.1). Here air in a flat wide stream is projected at a high velocity from an orifice well above the heads of the operatives in a path slightly inclined upwards. In this way a turbulence, agreeable as a cooling by fanning, is produced over a considerable distance; and it is frequently found that recirculating the factory air through the jet fans is better than introducing outside air through them. This system of personal cooling is suited to many industrial processes, where the operatives work in lines down the length of a large building. In one instance an air velocity of 200 ft/min was measured at head-level at a distance of 17 feet in front of the jet, and at the same time an air speed of 160 ft/min 62 feet distant. With temperatures above

70°F (21·1°C), and warm surroundings such fanning is much appreciated. Needless to say it would be more than unwelcome in a cold building.

The "ragged edge" of the current of entrained air surrounding the lower side of the upward-sloping main air jet, the initial velocity of which can be as high as 4000 ft/min, can be adjusted as the jet expands downwards with distance to give a desirable fanning velocity at head-level over a considerable floor distance. The degree of

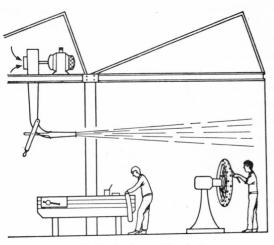

FIG. 7.1. Jet fan in a rubber factory.

cooling can thus be modified by altering the upward slope of the jet.

In one of H.M. bomb-filling factories during the 1939–45 war it was found that the men filling high explosives into bomb cases suffered from heat. Exhaust ventilating fans in the building roof were found to be unpracticable because they and their electric motors retained explosive dust. Small modified direct-driven propeller-type jet fans were therefore installed in the outside walls and impelled outside air along and over the positions of the operatives with excellent effect. It was found that not only were the men being cooled by the air movement, but also that, the particular factory building being rather small, these jet fans gave sufficient additional air change to make a marked improvement in an atmosphere that was essentially hot and laden with unpleasant and deleterious fume. An important feature of the arrangement was that the fans and their motors, always working in outside air, were free from deposits of dangerous dust.

HIGH TEMPERATURES IN BRITISH INDUSTRIES

In 1952 the Ministry of Labour and National Service published Welfare Pamphlet No. 5, 5th edition, *Heating and Ventilation in Factories*. This publication, now out of print, has not been reissued or revised since 1952, a fact which is, perhaps, significant. A fresh-air supply of from 400 to 550 cu. ft per hour per person is advised; with air movements for sedentary occupations of 30 to 60 ft/min velocity, not necessarily all of fresh air. The Factories Act of 1937 is quoted to the effect that no person in a workroom shall have less space around him than 400 cu. ft, exclusive of all spaces above a height of 14 feet from the floor. It is stated that the effective temperature scale, which takes account of the environmental factors, affords a convenient indication of what is a reasonable temperature, as required by the Factories Act. It also provides a useful method of analysing an unsatisfactory environment by indicating the main causes of the trouble, and shows how much improvement can be obtained by certain alterations in various factors. Examples are given. It may be noted that, although minimum allowable dry-bulb temperatures for workplaces are laid down in the Factories Act, and requirements for the control of wet-bulb temperature are laid down for certain industries in which process demands a certain high humidity in the air, there appears to be no corresponding requirement for maximum dry-bulb, or effective temperatures. In this respect the 1952 Pamphlet suggests that the Factory Inspectorate should refer to measurements of effective temperature, in deciding what constitutes a "reasonable temperature".

Summer Comfort Zone

Hickish (1955) made a number of surveys in tobacco factories, and also in the workrooms of Post Office workers. As a result, he drew up a hot-weather comfort zone applicable to such workers (Table 7.2). One of the notable aspects of his investigations was the conclusion that there is a critical discomfort point, which is marked by the occurrence of visible sweat found under the clothing.

TABLE 7.2. Factory Comfort Zone (Summer)

	Upper limits of comfort zone temperatures	Optima
Air dry-bulb	75°F 23·8°C	66·8°F 19°C
Globe thermometer	75°F 23·8°C	68·5°F 20·3°C
Equivalent temperature	73°F 22·8°C	65·9°F 18·7°C
Effective temperature	70°F 21·1°C	65·9°F 18·7°C
Corrected effective temperature	71°F 21·7°C	64·4°F 18·0°C
Dry kata-thermometer H	4·5	6·3

Cooling Methods

In this country the cooling of industrial buildings is generally effected by increasing ventilation; for which purpose the employment of the natural circulation of heated air, if properly controlled, is often found to be sufficient, provided that excessive process heat is removed or otherwise controlled. This should be done not only by venting hot air and fumes, but also by the screening of hot processes and machines to prevent the spreading of radiant heat into the premises and on to the operatives.

Bright metal screens or well-secured aluminium foil form good shields for radiant heat of all wavelengths. As previously noted, the prevention of unwanted insolation is important, and comparatively simple; but the growing practice of forming much of the wall area of new buildings, such as tall offices, of glass is likely to make the control of insolation more difficult.

REFERENCES

ANGUS, T. C. (1959) Air Conditions in Operating Theatres, Part 1. *The Medical Press*, **242**, 113, 134.
ANGUS, T. C. (1936) The Kata-Thermometer and Its Uses. *J. Inst. Heat. & Vent. Engrs.* **12**, 50.
HICKISH, D. E. (1955) Thermal Sensations of Workers in Light Industries in Summer: A Field Study in Southern England. *J. Hyg. (Camb.)* **53**, 112.
MARKHAM, S. F. (1947) *Climate and the Energy of Nations.* Oxford University Press.
MARKUS, T. A. (1964) Heat Transfer through Windows. Symposium on Glass in Modern Air Conditioning Practice. Privately printed: Pilkington Brothers.
Window Design and Solar Heat Gain (1966) *Building Research Station Digest* No. 68 (second series). H.M.S.O., London.

CHAPTER 8

TROPICAL HOUSING AND LIVING CONDITIONS

THE efforts of Westerners to aid the peoples of less-developed tropical countries to improve their dwellings and living conditions have not always met with success. It was reported by an Army Medical Officer serving in West Africa during the war that after new dwellings had been erected an African had asked him "Why do you take us from our own huts where the sun heat is kept off by our thatched roofs and bake us in these new buildings?" The buildings in question were roofed with flat sheets of dull-coloured iron, termed "pan-iron". Such a roof becomes extremely hot under a near-vertical sun and, in the absence of an inner ceiling, subjects the occupants to the effects of a large panel radiator.

Fry and Drew (1956), discussing the principles that should govern architectural design in the tropics, point out the differences between the type of protection human dwellings have to provide according to whether the prevailing temperature is below or above blood heat. In the first case, the main function of a house is to exclude cold and rain; in the second, it is necessary to provide for the body to lose heat at a rate dependent on the amount of work that it and its various organs have to perform. With ambient dry-bulb temperature near to or above blood-heat this essential cooling can only be accomplished by the evaporation of water from the body. In the hot, wet tropics, unless water vapour is extracted from the inside air, as by air conditioning, evaporative cooling of the body should be enhanced by air movement, artificial or otherwise provided, whilst as far as possible radiant heat should be excluded. In the hot, dry tropics, with low humidity, sweating is easy, and much water to drink is essential, whilst undue movement of hot dust-laden air should be avoided.

Fry and Drew give examples of how reasonable comfort can be obtained in tropical buildings without having recourse to expensive air-conditioning plants. One of these examples (Fig. 8.1) is that of a large library in Nigeria. In this library, as in other tropical buildings, where the solar heat is most fierce when the sun is almost overhead "sunbreakers" have been constructed. These are horizontal openings in the form of a honeycomb of precast concrete tubes passing through the walls and of considerable length. Radiant heat reaching the outer ends of these tubes, and at an acute angle, is dissipated to the exterior, and the sunlight which reaches the outer ends of these tubes is effectively diffused before entering the room. Overhanging horizontal shades or projecting flat roofs afford protection to walls and wall openings from a near-vertical sun.

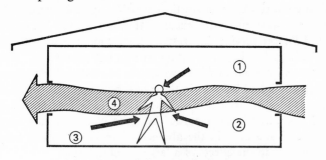

1, 2, 3 Heating by radiation.   4 Cooling by air movement

FIG. 8.2. Tropical architecture in the humid zone.
Maxwell Fry and Jane Drew.

Books are easily damaged by exposure to rapid changes of temperature and humidity. Before the building of the library was started complete air conditioning had been considered. It would have been possible to install the necessary machinery, but the cost would have been prohibitive. Another difficulty to be faced was the possibility of attack by certain insects that attack the spines of books. Success was obtained by providing cross-ventilation by natural air movement between window openings on either side of each room, protected against the entry of insects.

Figure 8.2 illustrates recommended principles for the design of smaller and simpler tropical buildings. The house is protected as far as possible from the heat of a near-vertical sun by the extended roof, which should have an open space between it and the interior ceilings, so that some of the absorbed and transmitted solar heat may be carried away by natural air movements. The outer surface of the roof should be whitened.

Large windows at about shoulder level allow movement of air

Fig. 8.1. The Library, University College, Abadan.

and this carries away sensible heat from building materials, with moisture emitted by occupants. In many tropical countries it is necessary to cover windows with some insect-proof gauze covering, especially at night; but the mesh should not be so fine as to cause too great a hindrance to air movement.

AIR CONDITIONING

By air conditioning is meant supplying air that is filtered free from airborne impurities and has the required temperature and humidity. Air velocity and direction at which the conditioned air enters the enclosure are also important.

Air conditioning would afford an ideal solution for the ventilation of all tropical buildings, but it is expensive, especially for large enclosures. Moreover, it can be said that for many tropical buildings air conditioning, although very pleasant, is somewhat of a luxury, and is seldom really essential. If it is installed, however, care must be taken to ensure that the cooling load and the capacity of the refrigerator should be properly worked out. In the humid tropics the load will include much more power to control the humidity to the desired value than in temperate climates because of the large amount of moisture that will generally have to be extracted from the fresh ventilation air drawn in.

EXAMPLES OF HUMAN REACTIONS
TO TEMPERATURE CHANGES

Angus and Brown (1957) described some experimental observations made on groups of students, many from overseas, in a London lecture room, which had been specially adapted so that ventilation rates could be varied and temperatures raised as desired. Many of these students were of western origin, but others came from India, Africa and from other hot countries; the latter group were, however, partly acclimatized to a London winter and appropriately clothed. As far as possible during these experiments the students pursued their studies and attended lectures normally; but at intervals they were asked to mark questionnaire papers to record their sensations in respect of heat, humidity, freshness or stuffiness. Figure 8.3 records how the two groups of students, from temperate and from tropical lands respectively, responded to changes in corrected effective temperature. From their replies to the questionnaire forms their subjective sensations were determined by the method described by Crowden and Lee (1940) which enables the summated sensations of heat or cold, humidity and freshness or stuffiness to be given numerical value.

The students of temperate habitat recorded, on average, that their most agreeable C.E.T. was between 65° and 66°F (18·3° to 18·9°C), whilst those from tropical lands preferred a C.E.T. between 64° and 68·2°F (17·8° to 20·1°C), with a peak for the greatest number of

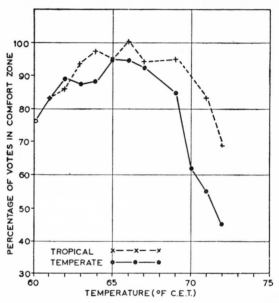

Fig. 8.3. The relation between summated subjective sensations of heat and moisture and C.E.T. (For 1261 votes, 852 temperate, 409 tropical subjects separately.)

comfort votes at 66·1°F (18·9°C). Figure 8.4 shows the results of another treatment of the same data. Here the physical parameter for comparison of subjects' sensations (according to Crowden's and Lee's criterion) is the dry-bulb temperature. For the "temperate" subjects the most popular dry-bulb temperature was 65·3°F (18·5°C), and for the "tropical" subjects the preferred dry-bulb temperatures ranged between 64° and 68·2°F (17·8° to 20·1°C), with a peak of the greatest number of comfort votes at about 65·1°F (18·4°C).

Because neither cooling nor dehumidifying was available, only varying degrees of air change and recirculation, it was not unusual for humidities to rise during the course of experiments. The observers noted that it was the students from the temperate rather than those from tropical climates who were the first to be conscious of any discomfort due to increases in humidity, apart from rises in temperature. This is reflected in the highest C.E.T.s recorded in Fig. 8.3, at 72°F (22·2°C) only about 44% of the "temperate" subjects were

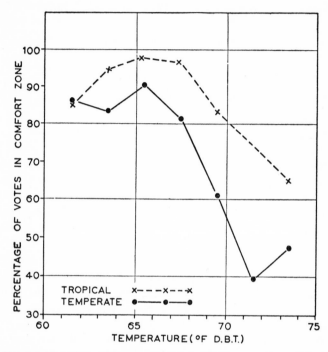

FIG. 8.4. The relation between summated subjective sensations of heat and moisture and D.B.T. (For tropical, 409 votes, and temperate, 852 votes, subjects separately.)

comfortable, as against some 69% of the "tropicals". It should be remembered that humidity plays a large part in effective temperature measurement.

TROPICAL COMFORT AND DISCOMFORT

Lee and Courtrice (1940) conducted a valuable examination of indoor climates found in tropical parts of Queensland, Australia, with accounts of the subjective impressions of the inhabitants, and

records of temperature observations. After examining other instrumental indices of thermal effects on man, the authors decided to use
effective temperature as their measure. They adopted Yaglou's
comfort zone for Americans (Yaglou *et al.*, 1932), which is as follows:

Season	No subjects feel comfortable below	50% subjects feel comfortable between	No subjects feel comfortable above
Winter	60°F	63 and 71°F	74°F
Summer	64°F	66 and 75°F	79°F

These authors noted the seasonal variations in effective temperature
with latitude in coastal areas, and in the arid hinterland. At Cape
York effective temperatures lie persistently above the comfort zone.
Passing southwards, the values depart more and more beyond the
comfort zone until, in winter, they come to lie below it. In Brisbane,
only the afternoon temperatures in the summer months lie above
the comfort zone. Referring to artificial fanning, Lee and Courtrice
wrote: "The cooling effect of a wind velocity of 150 feet per minute
(the level at which papers start to flutter and hence the desirable
limit of forced air movement) varies from 2°F E.T. on hot days to
3°F E.T. on cold days." This paper gives a table of day-degrees for
E.T. above and also below the comfort zone for 8 coastal, humid,
and 4 inland, arid places. The methods used appear to be admirable,
and may well be a model for the evaluation of climates in other
hot regions.

O'Dwyer (1950) gives a long discussion on physiological and
psychological effects of air conditioning in the tropics. (Since 1950
considerable developments have taken place in this field.) One
particular instance concerns a large factory in Africa with one
department of 75,000 cu. ft content. The point of particular interest
there is that the effects of air conditioning on personnel were quite
fortuitous. The air conditioning of this building, which was situated
practically in the jungle, was carried out entirely for the necessity
of process; the cost of the installation would have been too great
to justify it for any other purpose. The average number of persons
working in this department was usually about 100, rising to 200 on
occasion. The walls and roof of the factory were carefully insulated
and the air conditioning was intended only to reduce the relative

humidity to about 40 or 41% if possible. No attempt was made to lower the temperature in the factory because it was feared that African workers might suffer from the respiratory diseases to which they are prone if the temperature falls below 80°F (26·7°C); bad effects on the Europeans were also possible. The temperature outside the factory varied (in the month of February) from a minimum of 75°F (23·9°C) to a maximum of 95°F (35°C) at 3 p.m.; the relative humidity varied between 50 and 90%. With the air-conditioning plant in operation the maximum humidity indoors was reduced to 50% and the temperature remained fairly constant at about 80°F (26·7°C) to 85°F (29·4°C). This change in relative humidity, despite the small reduction in temperature, has proved a great advantage, both to the Africans and also to the Europeans. It was well expressed by one of the engineers who said: "I now work in a dry shirt and the difference in the amount of work I can get done was beyond my capability to understand before I worked here." On two occasions there was a failure of the plant and on each occasion not only did the Europeans suffer, both mentally and physically, from the change in environment, but there was a loud outcry from the African workers, who protested that it would be impossible for them to do the amount of work they were expected to do unless the air-conditioning plant was put in order. It is most interesting to note that, as far as is known, there has been no increase in respiratory disease among the Africans. O'Dwyer states: "There will be such a fall in efficiency if work is done in an effective temperature of 87·5°F (30·8°C) or over that on all counts the institution of air conditioning should be sound economics unless the building to be air conditioned is such that its adaptation would almost mean rebuilding." Another benefit of air conditioning, already mentioned, is the exclusion of flying insects from living quarters, with their menace to health.

COMFORT VOTES FOR SINGAPORE INTERIORS

Webb (1952) records the results of temperature surveys in Singapore dwellings and has calculated regressions† between degrees of thermal sensation on an eight point scale: (1) cold, (2) cool, (3)

† The regression equation and the regression coefficient of any two variables indicate if, and to what extent, one of these quantities changes in relation to the other. This relationship can only be valid when the points representing the

comfortably cool, (4) entirely comfortable, (5) comfortably warm, (6) warm, (7) hot, (8) excessively hot. Webb has denoted the comfort condition of satisfaction (number 4 on his scale) as C and related the values of C to effective temperature $T(°F)$ as $C = 0·81T − 59·8$. Table 8.1 records the results of Webb's temperature surveys, and the thermal sensations of the occupants of a variety of Singapore dwellings; it should be remembered that C was found to represent a very desirable thermal sensation. Singapore is in latitude 1° 20′N. Owing to the equability of the climate (the annual daily range of mean temperature being only 20% of the mean daily range) and its warmth, sensitivity after long residence might be expected to be high. The dwellings ranged from traditional *atap* (thatch) roofed bungalows, brick and tile houses and bungalows with metal roofs, to a motor vehicle standing in the open. In all fourteen premises were investigated. Variations in the dress of the observers, who acted as subjects, were not great. Typical day dress was of cotton and consisted of short-sleeved shirt, trousers, underpants and shoes, with socks and necktie optional. Radiation indoors was usually negligible. The difference between dry-bulb and globe thermometer was usually less than $\frac{1}{2}°F$. Hence the effective temperature was not corrected.

Table 8.1 shows how comfort votes on Webb's scale were distributed in terms of effective temperature among the subjects in the fourteen premises, and shows that the most popular effective temperature was 79·2°F (26·3°C).

Webb concludes that the warm period at midday could be made more comfortable by increasing the indoor air velocity from the average of less than 100 to about 200 ft/min. At night very low air velocities are common; and the house should be as open as possible, though violent storms should be guarded against.

---

changes in values lie on or nearly on a straight line. In Webb's observations the changes in thermal sensation on an eight point scale are related to the changes of effective temperature in Singapore dwellings.

A coefficient of correlation is a useful measure of the degree of association between two variable quantities. These coefficients may be either positive if the dependent variable increases with the independent variable, or negative if the reverse is true. Coefficients of correlation of +1 or −1 show a complete agreement in simultaneous variations. Values of less than unity show that agreement is imperfect. In the example given by Ho (page 93) the vapour pressure has a higher partial coefficient of correlation than either air dry-bulb temperature or air velocity and therefore a greater influence on human thermal sensation.

TABLE 8.1

Comfort vote	C	Mean effective temperature		Number of obser- vations	Percentage of votes	
		°F	°C		C = 4	C = 3, 4 or 5
Cool	1½–2½	76·5	24·7	27	11	52
Comfortably cool	2½–3½	77·8	25·4	44	23	63
Comfortable	3½–4½	79·2	26·3	90	19	70
Comfortably warm	4½–5½	80·4	26·9	135	16	67
Warm	5½–6½	81·7	27·6	75	21	65
Hot	6½–7½	83·0	28·3	33	3	24

Ho (1952) extends the data given by Webb, and gives the following partial correlation coefficients of the three main physical variables with C, the comfort values on Webb's scale.

	Partial correlation coefficient
Air temperature (dry-bulb)	0·52
Vapour pressure	0·53
Air velocity	0·44

Ho establishes an empirical equation

$$C = 0·33t + 0·48p + 0·22v^{\frac{1}{2}} - 33·4,$$

where C = comfort vote,

t = dry-bulb temperature, °F,

p = vapour pressure, mm mercury,

v = air velocity, ft/min.

The total correlation of the comfort condition C with the three factors taken together is found to be 0·67, which is appreciably higher than that derived with mean effective temperature, and may well justify its use as the basis of a Malayan comfort scale.

Yap Tien Beng (1956) gives figures from observations taken in Singapore buildings in the afternoon (the warmest period of the day). The object of the investigation was to give a guide to the construction of new dwellings. Thirteen different types of houses were investigated, by the approved instrumentation. The indoor air velocities cited were found to vary between 16 and 226 ft/min. Where ceiling fans were installed these were not running during tests. One notable fact is that in dwellings with ceilings under roofs the corresponding E.T. differed very little inside the building from that outside. It is not stated whether the sky was clouded, or otherwise. In one Nissen hut with no inner ceiling the inside D.B. was 89·2°F (31·8°C), the air velocity 20 ft/min, and the C.E.T. 86°F (30°C). Outside the corresponding values were: D.B. 85·0°F (29·4°C), air speed 100 ft/min, and C.E.T. 81°F (27·2°C). This shows that in quest of cool dwellings double roofing is desirable; also large windows, suitably screened. (In the standard Nissen hut there are no side windows.) Yap Tien Beng found no significant difference in internal conditions between flats of low and normal ceiling height.

RECORDED TEMPERATURES IN OTHER HOT DISTRICTS

Leithead and Lind (1964) give a hygrometric chart Fig. 8.5 on which are plotted the midday outside dry-bulb and wet-bulb temperatures for various desert and humid tropical localities during the summer months. Also shown are the upper limits of climatic conditions to be found in South African mines (S.A.) and Indian Gold Mines (I), in coal mines in Great Britain (B) and in the Ruhr (R), and also in armoured tanks in the desert (T). It is of interest to compare, in terms of effective temperature, the outdoor conditions at some of the localities for which data are given in Fig. 8.5, where, despite the great heat, particularly by European standards, men live and work throughout the year.

Table 8.2 gives the dry-bulb and wet-bulb temperatures and the relative humidity recorded at eleven of these places. In the last columns the effective temperatures corresponding to indoor conditions with these outside D.B. and W.B. temperatures have been extracted from Fig. 2.6 (p. 19). Effective temperatures with air velocity of 20 ft/min are given in columns 7 and 8, representing the E.T. that may be expected in a nearly closed room. To indicate the physiological cooling that might be expected with mechanical fanning capable of increasing the rate of air movement to 200 ft/min the reduced effective temperatures so obtained are given in columns 9 and 10; whilst in columns 11 and 12 are given the corresponding reductions in E.T., as differences. Inspection of the figures in the last two columns indicates that increased air movement has greater

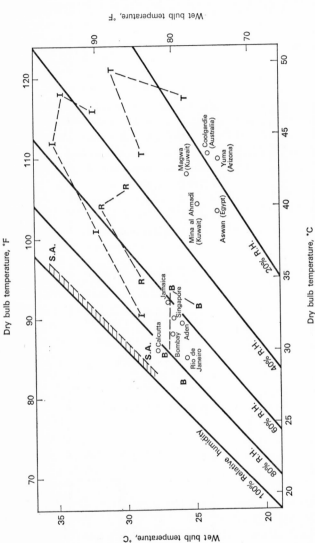

FIG. 8.5. A psychrometric chart showing typical midday dry- and wet-bulb temperatures in various desert and tropical localities during summer months; the values shown may vary from season to season and do not take solar radiant heat into account. Also shown, for comparison, are the upper limits of climatic conditions to be found in South African (S.A.) and Indian (I) gold mines, in coal mines in Great Britain (B) and in the Ruhr (R), and in armoured tanks (T) in the desert.

TABLE 8.2

| | Air temperatures | | | | Relative humidity % | Effective temperatures with air velocities of | | | | Reductions in effective temperatures with increased air velocities | |
| | Dry-bulb | | Wet-bulb | | | 20 ft/min | | 200 ft/min | | | |
	°F	°C	°F	°C		°F	°C	°F	°C	°F	°C
Aden	90	32	79	26	60	83·2	28·4	81·4	27·4	1·8	1·0
Aswan	104	40	74	23	27	84·8	30·6	83·8	30·0	1·0	0·5
Bombay	88	31	81	27	73	83·5	28·6	81·5	27·5	2·0	1·1
Calcutta	86	30	81	27	80	83·0	28·3	80·8	27·1	2·2	1·2
Coolgardie	111	44	75	24	18	86·7	30·4	85·9	29·9	0·8	0·4
Jamaica	91	33	81	27	60	84·6	29·2	82·7	28·2	1·9	1·1
Magwa (Kuwait)	108	42	79	26	27	87·8	31·0	86·9	30·5	0·9	0·5
Mina al Ahmadi (Kuwait)	104	40	75	24	30	85·2	29·6	84·3	29·0	0·9	0·5
Yuma, Arizona	110	43	73	23	18	85·6	29·9	84·9	29·4	0·7	0·4
Rio de Janeiro	84	29	77	25	72	80·0	26·7	77·8	25·6	2·2	1·2
Singapore	92	33	81	27	65	84·9	29·4	83·2	28·5	1·7	0·9

cooling value in humid than in dry heat, but it is realized that the data presented are too few to allow of statistical analysis.

AIR CONDITIONING OF DWELLINGS

Grocott (1948) has given a comprehensive account of large-scale air conditioning which has been applied sucessfully in the living quarters of Europeans working in the Persian Gulf. These installations have been in use for a number of years, and ample data have been recorded for desirable indoor conditions in this particular locality. The air-conditioning installations described by Grocott are on a very large scale. He gives the plan for the supply of chilled water in underground mains to an estate of considerable size. The object of this scheme is to enable separate houses to have their own air-conditioning sets drawing their "cold" from mains; even as heat is drawn from mains for the houses in district-heating estates in cold countries. This chilled water enables people to cool and dry the air in their houses as desired.

Grocott sums up the findings of much experience as follows:

TABLE 8.3. Desirable Conditions (Persian Gulf)

(a) Maximum inside temperatures: 85°F (29°C) dry-bulb.
　　　　　　　　　　　　　　　　　77·5°F (25·3°C) effective.

(b) Inside summer ranges: 75–83°F (23·9–28·3°C) dry-bulb.
　　　　　　　　　　　　　　70–77°F (21·1–25°C) effective.

(c) Relative humidity shall be a maximum of 60%.

(d) Relative humidity may cycle between 25% and 50% providing the dry-bulb temperature is right.

(e) The dry-bulb temperature may cycle through about 6°F (3·3°C) without being noticeable.

(f) For winter comfort a dry-bulb temperature of 70°F (21·1°C), with relative humidity as required by (d) is acceptable.

Grocott continues: "It is dangerous to generalize but it seems reasonable to suppose that (c), (d) and (e) are applicable in every humid tropic."

Single-room Air Conditioning

Crowden and Angus (1938) described experiments in artificially produced tropical conditions in London. A large air-conditioned room was available in which temperatures and humidities of tropical severity were produced. Within this room was erected an experimental cubicle large enough for bed, table and chair; its dimensions were 10 ft by 7 ft and 6 ft 2 in high. Such a cubicle could provide the manager of a mine in a very trying climate with a small, comfortable, air-conditioned office where he could also obtain a good night's rest. The cubicle was constructed of panels of light building board held in aluminium framing. Each panel was composed of two sheets of building board with an air space of about one inch between them. This space was divided into two by a diaphragm of aluminium foil; the whole being heat insulating and moisture proof. These prefabricated panels were provided with clips by means of which the whole cubicle could be quickly erected by unskilled labour.

Such cubicles could be exported to tropical lands and there erected in any part of an existing house. The cost of the whole, with a small console containing a suitable air-conditioning plant, would be far less than the cost of air conditioning a whole house. In the experimental cubicle, the air-conditioning plant cooled, dried and recirculated the air in the cubicle, to which was added a small proportion of air taken from the heated and humid air of the enclosing large air-conditioned room. With two men slightly active in the cubicle, the following results were obtained:

TABLE 8.4

	Outside air temperatures		Inside cubicle		Supply air from plant	
	°F	°C	°F	°C	°F	°C
Dry-bulb	96	35·6	81	27·2	63	17·2
Wet-bulb	90	32·2	70	21·1	59·5	15·3
Dew point	88·5	20·3	65	18·3	57·5	14·4
Relative humidity	79%		58%		82%	

During this test 20 cu ft/min of the surrounding air were being introduced through the air conditioner. In this and in other similar experiments the power required was less than one kilowatt.

CONCLUSIONS

In those tropical lands where heat and especially humidity are so excessive that many Europeans experience considerable distress it has been shown that when careful architectural design fails to give protection, air conditioning is much appreciated. This is essentially a costly measure, but from much evidence, some of which has been quoted, one may ask whether, in regions where insolation is not the main cause of thermal stress, but where high humidity with high, but not extreme, temperature prevails, the need to devote great attention to the heat insulation of roofs and walls is as great as has been supposed. In such situations it would seem that a large dry-bulb depression below the outside temperature is not important, and may even be undesirable. This raises an interesting question. When air conditioning is to be provided in a building situated in a humid tropic, we know that the greater part of the cooling load will probably be utilized in the removal of water vapour from the entering outside air. In view of this, should extra precaution be taken to guard against more unwanted moisture entering by diffusion through the materials of which the house is built, and is the quantity of water so entering large enough to be of engineering significance in the overall design? It is well known that many building materials, even if "air-tight", freely transmit moisture by diffusion, or capillary action; see the "damp courses" in the lower parts of the brick walls in our own houses. In the refrigerating industry, where the preservation of food is the object, great value is placed on water-vapour barriers, perhaps in the form of membranes in the walls and other enclosures of refrigerated spaces. So, in dwellings in the humid tropics will it be good policy to provide such water barriers in the walls, roofs, and perhaps floors of air-conditioned buildings?

Here, perhaps, is a little-explored field of research for the architect, the physicist and the engineer.

On the other hand it has not been possible to find evidence of complaint that air-conditioning plant produces too much drying in humid tropics. As has been stated Africans like this feature of air conditioning, though they are very susceptible to low dry-bulb temperatures. In Northern Nigeria, where the dry-bulb temperatures are generally in the 90°F (32½°C) range, with low humidity, most Europeans are perfectly comfortable when wearing the usual very light clothing. But it has been stated that on the rare occasions when the temperature falls into the region 60°F (15½°C), Africans become sluggish and find work difficult. Usually in Great Britain after an unusually hot summer our own people are stimulated to greater activity when heat gives place to autumn coolness. Such essential differences should not be looked upon as any kind of foolish racial prejudice, or discrimination, but as facts that should be borne in mind by those responsible for the planning of work and living places.

REFERENCES

ANGUS, T. C. and BROWN, J. R. (1957) Thermal Comfort in the Lecture Room: An Experimental Study of Winter Requirements. *J. Inst. Heat. & Vent. Engrs.* 25, 175.

CROWDEN, G. P. and LEE, W. Y. (1940) Sensations of Heat and Moisture. *Chin. J. Physiol.* 15, 475.

CROWDEN, G. P. and ANGUS, T. C. (1938) The Control of Indoor Climate by Air Conditioning with Special Reference to the Tropics. *J. Inst. Heat. & Vent. Engrs.* 6, 442.

FRY, MAXWELL and DREW, JANE (1956) *Tropical Architecture in the Humid Zone.* Batsford, B. T., London.

GROCOTT, J. F. L. (1948) Comfort Cooling in the Tropics. *J. Inst. Heat. & Vent. Engrs.* 16, 36.

HO, P. Y. (1952) Correlations of Equatorial Climatic Factors with Comfort. *J. Inst. Heat. & Vent. Engrs.* 20, 196.

LEE, D. H. K. and COURTRICE, R. (1940) Assessment of Tropical Climates in Relation to Human Habitation. *Trans. Roy. Soc. Trop. Med. & Hyg.* 33, 601.

LEITHEAD, C. S. and LIND, A. R. (1964) *Heat Stress and Heat Disorders.* Cassell, London.

O'DWYER, J. J. (1950) On Some Aspects of Air Conditioning in the Tropics (Napier Shaw Premium Lecture). *J. Inst. Heat. & Vent. Engrs.* 18, 84.

WEBB, C. G. (1952) On Some Observations of Indoor Climate in Malaya. *J. Inst. Heat. & Vent. Engrs.* 20, 189, 289.

YAGLOU, C. P., CARRIER, W. H., HILL, E. V., HOUGHTEN, F. C. and WALKER, J. H. (1932) How to use the Effective Temperature Index and Comfort Charts. *Trans. Amer. Soc. Heat. & Vent. Engrs.* 38, 410.

YAP TIEN BENG (1956) Day Time Corrected Effective Temperatures inside Buildings in Singapore. *Med. J. Malaya,* 10, 326.

APPENDIX†

THE CONTROL OF RE-RADIATED SOLAR HEAT
BY WHITEWASH

Solar heat, re-radiated at night, from the inside surfaces of walls and ceilings materially adds to thermal discomfort caused by the high humidity of the air in the Yaba district near Lagos. Readings taken in March 1945 showed that at 8 p.m. the temperature of the inner surface of the nine inch sandcrete and plaster SW. wall of a bungalow was frequently 91°F or more when the indoor air temperatures were 85–86°F dry bulb and 80°F wet bulb.

Photographs, tables and graphs of inside surface temperatures taken in tests with and without whitewash on the external surface of the bungalow wall showed that, as a direct result of reflection by whitewash in preventing the absorption of a large proportion of solar radiation during the day, the surface temperature of the inner side of the wall was reduced by 3 or 4°F or more at night.

A reduction of wall and ceiling temperature to even a few degrees below 90°F is of particular advantage in lessening thermal discomfort under a mosquito net, which measurements have shown to reduce air movement from 80 to 10 feet per minute.

The following table shows the results of tests with two corrugated pan iron roofing sheets one of which was whitewashed on its upper surface. The whitewash was prepared and applied according to the directions given for U.S.A. "Government Whitewash" as used on lighthouses.

UNDER-SURFACE TEMPERATURES OF CORRUGATED IRON
SHEETS WITH WHITEWASHED AND PLAIN OUTER SURFACES
EXPOSED TO TROPICAL SUN,

LAGOS, March 17, 1945

Time	Shade temperatures outdoors		Temperatures of under surface of corrugated iron sheets		Under surface temperature difference
Hrs	Dry bulb °F,	Wet bulb °F,	Whitewashed °F,	Plain °F,	°F
1440	85·5	79·5	106·0	127·0	21·0
1445			108·5	134·0	25·5
1550	90·1	81·0	106·5	128·0	21·5
1630			99·0	114·0	15·0
1725	87·0	80·2	93·5	102·5	9·0
1810	85·5	80·0	86·5	89·0	2·5
1833	84·5	80·0	84·5	85·0	0·5

† Extract from Crowden, G. P., *Transactions of the Royal Society of Tropical Medicine and Hygiene.* **40**, 1947, 362p4.

By thermometers adhered to the reverse surfaces of whitewashed and blackened strips of sheet iron, similarly exposed to light from an electric arc, the reflecting properties of whitewash for short wavelength radiation were demonstrated. A "Leslie Cube", Moll Radiation Thermopile and Galvanometer were used to demonstrate the fact that, at indoor surface temperatures commonly encountered in the Tropics, a white surface emits as much radiant heat of long wavelength as a black surface. It followed from this that no advantage to thermal comfort would be gained by whitewashing interiors, and that its value for this purpose depended on its use on external surfaces for reflecting direct solar radiation.

DURABLE WHITEWASH

NOTE:

The following general directions and Recipe 1 for Durable Whitewash are taken from Extension Pamphlet, Helpful Hints Series No. 1, November, 1939, by Kenneth H. Prior, C.M.S. College, Agricultural Department, Awka, Nigeria.

General Directions

It is in response to a number of requests for a serviceable whitewash that this leaflet has been produced and it is hoped that it will prove useful. The information given is gleaned from reliable sources.

(i) *Value of Whitewash.* Whitewash is worthy of a much more popular place in West Africa than it enjoys at present. It can be used for a multitude of purposes and when a durable formula is used it can well be considered as a close rival of paint and yet much cheaper. Its ingredients besides being cheap are usually readily available. It is not difficult to make and it is easy to apply. The chief uses of whitewash are: to brighten dark interiors, to cover stained and unsightly areas, to mark dangerous corners and objects, to preserve exposed surfaces from the weather and also to lower temperatures. A white surface reflects heat and results in lower interior temperatures. A durable whitewash could be used with distinct advantage on a good many of the "pan" roofed buildings in the country.

For those who like a little colour on interior walls, rather than a "dead" white, colouring matter whose pigments are not affected by lime can be added. Among such are yellow ochre and raw and burnt sienna. For sanitary purposes a disinfectant can be added.

(ii) *Preparing the Surface.* Contrary to common belief the surface for whitewashing needs to be as thoroughly prepared as for painting. The surface should be scraped if rough and then brushed to remove all dust. If the surface is dampened just previous to whitewashing it will help considerably.

(iii) *Applying the Whitewash.* The work should be done quickly and as evenly as possible, but do not attempt to brush out as in painting. A wide brush should be used or a sprayer if large areas are to be attempted. When using a sprayer the nozzle should be kept clean or the stream will become uneven and the work patchy.

(iv) *Estimating Quantities Required and Time Needed.* It is estimated that a general basis for work is: 1 gallon of whitewash for 225 square feet of wood surface, or 180 square feet of brick surface or 270 square feet of plaster surface. A man using a 4 inch brush can cover 200 square feet of ceiling, or 200 square feet of rough walls, or 350 square feet of smooth wall in one hour.

(v) *Preparing the Wash: Directions for Slaking Quick Lime.* Water-slaked *quick* lime is customarily used and care must be exercised in the slaking of it. It is better done with boiling water, but cold water can be used. Add the water, a little at a time, to the lime in a clean wooden bucket, keg or barrel. When slaking is well started add more water gradually to replace that lost in the slaking process. If not enough water is added the lime will become "scorched" and part of it will be granular. If, on the other hand, too much water is added at one time it may retard or "quench" the slaking process. After the lime is completely slaked add enough water to make a thick paste, cover the container and allow to stand for several hours or overnight.

RECIPES FOR DURABLE WHITEWASH

Recipe No. 1

FORMULA (for HIGH GRADE work). This whitewash is known in the U.S.A. as "Government Whitewash" as it is used on lighthouses and other Government buildings. It is durable and even though exposed to the weather, does not peel off.

Take 38 lbs of good unslaked lime; slake with boiling water, covering during the process to keep the steam in; strain the liquid through a sieve fine enough to retain all unslaked lumps. Dissolve a pack (1 lb) of clean salt in a little water and add to the solution; boil to a thin paste three pounds of rice and put the paste into the mixture while still hot, add one pound of glue, previously melted over a fire, and one-half pound of whiting. Mix well and then add five gallons of hot water, stirring well: cover closely and let stand for several days.

Colouring matter may be used, varying the tint to suit the taste.

No matter what quantity is desired, these are the proportions in which the ingredients are used. If whitewash can be applied hot it will be better and last longer.

N.B. 1. The whiting can be omitted.

2. 38 lbs of lime require approximately 6 gallons of water for slaking.

Recipe No. 2

Place $\frac{1}{2}$ bushel (20 lbs) of fresh quick lime in a barrel and slake with warm water. When boiling point is reached add about 5 lbs Russian tallow and stir in well. Cover the barrel to conserve the heat.

When cold, strain and use in consistency of cream.

Recipe No. 3

Limewash. 14 lbs quick lime and $\frac{1}{4}$ lb common salt. Slake in approximately $1\frac{1}{2}$ gallons warm water. At boiling point add 1 lb tallow or 1 pint boiled linseed oil and stir well. Strain and use as above.

Recipe No. 4

"Snow-crete" Whitewash. Snow-crete mixed with water to form a whitewash can also be used effectively.

NAME INDEX

SUBJECT INDEX

107